ZERO to MAKER

*Learn (Just Enough) to Make
(Just About) Anything*

David Lang

MAKER **MEDIA**
SEBASTOPOL, CA

ZERO TO MAKER

by David Lang

Copyright © 2013 David Lang. All rights reserved.

Printed in the United States of America.

Published by Maker Media, Inc., 1005 Gravenstein Highway North, Sebastopol, CA 95472.

Maker Media books may be purchased for educational, business, or sales promotional use. Online editions are also available for most titles (*http://my.safaribooksonline.com*). For more information, contact O'Reilly Media's corporate/institutional sales department: 800-998-9938 or *corporate@oreilly.com*.

Editor: Brian Jepson	**Cover Designer:** Jason Babler
Production Editor: Kara Ebrahim	**Interior Designer:** Monica Kamsvaag
Copyeditor: Dianne Russell, Octal Publishing, Inc.	**Illustrator:** Rebecca Demarest
	Cover Illustrator: Nate Van Dyke
Proofreader: Elise Morrison	**Cover Photographer:** Gunther Kirsch
Indexer: Judy McConville	

September 2013:   First Edition

Revision History for the First Edition:

    2013-08-22:   First release
    2014-01-17:   Second release

See *http://oreilly.com/catalog/errata.csp?isbn=9781449356439* for release details.

ISBN: 978-1-449-35643-9
[LSI]

For my mom and dad. Where they found the patience, I'll never know.

Contents

Preface

Safari® Books Online

Note

Safari Books Online is an on-demand digital library that delivers expert content in both book and video form from the world's leading authors in technology and business.

Technology professionals, software developers, web designers, and business and creative professionals use Safari Books Online as their primary resource for research, problem solving, learning, and certification training.

Safari Books Online offers a range of product mixes and pricing programs for organizations, government agencies, and individuals. Subscribers have access to thousands of books, training videos, and prepublication manuscripts in one fully searchable database from publishers like MAKE, O'Reilly Media, Prentice Hall Professional, Addison-Wesley Professional, Microsoft Press, Sams, Que, Peachpit Press, Focal Press, Cisco Press, John Wiley & Sons, Syngress, Morgan Kaufmann, IBM Redbooks, Packt, Adobe Press, FT Press, Apress, Manning, New Riders, McGraw-Hill, Jones & Bartlett, Course Technology, and dozens more. For more information about Safari Books Online, please visit us online.

How to Contact Us

Please address comments and questions concerning this book to the publisher:

Maker Media, Inc.
1005 Gravenstein Highway North
Sebastopol, CA 95472
800-998-9938 (in the United States or Canada)
707-829-0515 (international or local)
707-829-0104 (fax)

We have a web page for this book, where we list errata, examples, and any additional information. You can access this page at *http://oreil.ly/zero-maker*.

To comment or ask technical questions about this book, send email to *book questions@oreilly.com*.

Maker Media is devoted entirely to the growing community of resourceful people who believe that if you can imagine it, you can make it. Maker Media encourages the Do-It-Yourself mentality by providing creative inspiration and instruction.

For more information about our publications, events, and products, see our website at *http://makermedia.com*.

Find us on Facebook: *https://www.facebook.com/makemagazine*

Follow us on Twitter: *https://twitter.com/make*

Watch us on YouTube: *http://www.youtube.com/makemagazine*

Acknowledgments

In true DIT spirit, this book was made possible thanks to the love, support, and generosity of dozens of people.

I'm eternally grateful to the team at Make, who gave this project a chance: Dale Dougherty, Gareth Branwyn, Brian Jepson (my intrepid editor), Mark Fraunfelder, Sherry Huss, and everyone else. The team at TechShop that took me under their wing: Mark Hatch, Jim Newton, Dan Gonzalez, Zack Johnson; and all the makers mentioned in this book who put up with me, especially Tim Anderson.

Writing-wise, I never would have made it without the patience and encouragement of Ethan Watters, Jason Rezaian, Stephen Hanselman, and Julia Serebrinsky. And everyone who read drafts and provided such wonderful feedback: Malcolm Knapp, Abe Fetterman, and Mom!

Writing this book served as a constant reminder of how lucky I am to know and work with Eric Stackpole (and everyone else in the OpenROV community). What a crazy ride. I'm looking forward to the many more adventures to come.

And, of course, all 2,582 Kickstarter backers. Writing is a lonely battle, and their support was wind in the sails. A special thank you to the True Believers: Adam Anderson, Aimee Sonderman, Alan McNeil, Alexander Kiselev, Andrew Buckman, Andrew Foster, Andrew Thaler, Anton Willis, Antony Evans, Bhushan Lele, Brad Doane, Brett Bayley, Brian Boyle, Charlotte Zejlon, Chris Lin, Christian Latouche, Colin Ho, Courtney Harrison, Craig Selvaggi, Cyril Ebersweiler, Daemon Eye, David Cooper, David Lusby, Davis, Deepak Mehta, Dominik Fretz, Douglas Schuberth, Douglas Yee, Emily Pilloton, Eric Facas, Eric Hall, Ezequiel Calderara, Fadi Musleh, Frank, Fred & Bonnie davis, Friedrich Boeckh, Gillian Benary, Giovanni Farinazzo, Greg P Flanagan, Gretchen Frankenthal, Jacob Hurwitz and David Neill,

Jacque Pruitt II, James, James Green, JamesLechner, Jason Lucas, Jay, Jeff Konrardy, Jeff Petre, Jeff Ziehler, Jeffrey Corbett, Jeremiah Rogers, Jessica Jackley, Jocelyn Corbett, Joe Lillibridge, John La Puma, John Parts Taylor, Julia Kernitz, Justin Shaw, Kacy Oen, Karen Baumgartner, Kate Linge, Katherine Adams, Katie Wilson, Keith Chapman, Keith Nelson, Keith Woeltje, Kevin I, Kiel Luse, Kim, Kosi2801, kwyjibo, Kyle Smith, Lawrence Neeley, Lisa Ballard, Lisa Q. Fetterman, Lynda Davis, MakerStash, Mark Pereyda, Matthew F. Reyes, Matthias E., Max Fancher, Melissa Spencer, Michael Pepper, Nathan, ND356, Nick Campbell, Nick Pinkston, Nicole Tindall, Nik Martelaro, Olivier Vigneresse, OttawaGuy, Paolo Tabaroni, Paul Lang, Rachel Tobias, Richard Brull, rkt88edmo, Ronny, Ryan Huff, Sabrina Merlo, Sam Brown, Samantha Farbman, shawn looker, Stephen Marchand, Steve Huynh, Steven Keating, Sue Lang, T.J. Anderson, Tamara Dunaye, Teresa Gonczy, Thanh Vo, Tom Lang, Tom Waloszek, Tony Guntharp, Tucker Max, vlaskovits, Yasser Ansari, and Zach Berke.

Down the Rabbit Hole

The cave you fear to enter holds the treasure you seek.

— JOSEPH CAMPBELL

The entire situation was unfamiliar. I was in a part of the world I had never seen —the foothills of the Trinity Alps in Northern California, deep in the heart of Humboldt County, where the cell phone reception seemed as prehistoric as the surrounding landscape. We completed the seven-hour drive from San Francisco, through the giant redwood forest, to explore the Hall City Cave. The beam of light emanating from my headlamp exposed vivid details: threatening stalactites, a rock wall covered in spiders, and a few inverted, sleeping bats. I was carrying a large yellow Pelican case containing an underwater robot I had helped design and build. That was really new.

"I think the next time we do this, we should wait until the summer," I said to Eric as I handed him the case to get a better footing as we descended further into the cave. The clunky, waterproof boots I was wearing were not the best choice for spelunking, but they were my only option given the awful weather outside the cave. He looked at me and smiled. It was obvious to all six brave souls who made the trek that choosing a January date for our trip to the cave was not wise. With such a dry and mild winter we thought Mother Nature might spare us a few more nice days, but we had pushed our luck. The heavy, constant snowfall was an hourglass constantly reminding us of how little daylight was left and how much worse the return trip could get.

My remark to Eric was meant to be lighthearted. A series of nearly trip-ending incidents had left the group exhausted. We woke up to worse-than-expected weather and were forced to scramble to find chains for our car tires. After we made it up the mountain, we found the back roads to be impenetrable. Luckily, we met a local Wildwood resident who offered to help plow us through the snow-covered back roads. By the time we reached the cave, everyone was tense and tired.

Ever since Eric Stackpole and I first met and talked about underwater Remotely Operated Vehicles (ROVs) and ocean exploration, we had been plotting for this moment. Eric tells the backstory, The Legend of Hall City Cave, much better than I do; it's his favorite story to tell and he will share it with anyone willing to listen, whether it's a full auditorium or a small dinner party. He always starts out the same way: ensuring that his audience has a full 90-second attention span to dedicate to his tale. He shakes out his arms and takes a deep, preparatory breath, "Whooooosh!" A wave of his hands and an exaggerated exhale take them back in time:

> *Flashback, 1800s. Northern California. Gold rush. Two Native American men rob a gold mining operation and make away with an estimated 100 pounds of gold. A sheriff's posse is assembled to track down the men. After days of chase, they eventually catch the two men, but they no longer have the gold. The sheriff's posse makes an offer, "Tell us where you hid the gold and we'll spare your lives." The men explain that they hid the gold in the Hall City Cave. Despite the sheriff's promise, both men are hung on the spot. The posse returns to the area the men described and, sure enough, there's a cave. They don't find the gold, but toward the back of the cave they find a hole six feet in diameter and filled with water. The underwater cavern extends down further than they can see, and they lack the tools or technology to explore further, so the sheriff's posse gives up.*

Eric ends the story by recounting the numerous cave divers and treasure hunters[1] who chased the legend as far as a human diver could possibly explore, without ever finding the bottom. His final line is "and that's why we're building this underwater robot: to solve the mystery of the Hall City Cave." More information on the story of Hall City Cave is shown in Figure 1-1.

I've heard the story a hundred times, and it never gets old. When Eric and I first met, that's really all it was: a great story and a rough prototype of a robot he wanted to build. Even though I didn't have any relevant technical experience, I knew I wanted to be a part of the adventure. The idea of exploring the unknown with an ingenious tool made from off-the-shelf parts held me in its grip. It struck a chord inside me that my office job couldn't possibly reach.

---

1. Eric originally heard about the cave from a friend and got most of his initial information from Dave McCracken's website (*http://www.goldgold.com/cave-diving.html*).

Figure 1-1. News of the cave

Now, as I walked inside the cave with Eric and the ROV, I could hear my own heartbeat. About 15 meters inside the cave, descending its rocky steps and twisting caverns, part of me still didn't believe the underwater hole was real; could it be that this was just an urban legend to lure tourists into the Wildwood Store just a few miles away? Part of me began to doubt the whole thing. But as we came upon what seemed to be the end of the cave, we flashed our lights toward the floor and there it was: a hole six feet wide, filled with crystal-clear water deeper than the flashlight could illuminate. Just as the story told.

Eric set down the Pelican case near the underwater opening and took out the ROV. The combined direction of our headlamps lit up the work area, as shown in Figure 1-2. As we inspected the robot, Eric noticed the walk to the cave had caused one of the propeller ducts, the circular guards that wrap around the propeller, to crack off—not ideal, but not critically important. We decided to break the opposing duct to make it proportionate.

Figure 1-2. The work area

Even without the propeller ducts, the robot is a beautiful piece of bare-necessity engineering. Well, at least beautiful once you know what you're looking at. Like most underwater creatures and contraptions, it looks awkward out of its natural habitat. The brain of the ROV, the main electrical system, is housed in a clear plastic cylinder that includes the camera, three speed controllers, and the micro-controller. The cylinder resembles a large French press, except horizontal, filled with electronics and built to withstand pressure. It's kept air-tight with plastic end-caps. Wires and communication lines protrude from the end-caps and are potted with epoxy. In addition to keeping the electronics dry, the cylinder also serves as the main force of positive buoyancy to keep the ROV upright underwater. The outer shell of the ROV is a sheet of blue acrylic plastic that folds tightly over the cylinder, like downward folded wings. The wings extend down to the battery packs—six C batteries, three on each side—that also act as ballast to counteract the air-tight container. Add in the motors, propellers, and a few threaded steel rods, and the entire robot is still only about the size of a small microwave.

Eric started in with the last minute waterproofing, while I added weights to the steel rods to make sure that the robot had the correct buoyancy. Brian Lam, our photographer, arranged flashlights and made sure the cameras were rolling. Jeff Bernard and Bran Sorem, friends who had decided to join us for the trip, maneuvered themselves along the wall of the cave in order to shine lights into the cavernous depths. Zack Johnson, another friend and robot collaborator, unwound the tether, which would be the communication line to the robot when it was underwater. The moment of truth was finally here.

Eric walked over and set the robot into the water. It floated on the surface and we collectively held our breath and waited for the robot's next move. The LED lights switched on, like an infant opening its eyes. The silence was broken by the buzzing of the robot's propellers. Sporadic at first, it took several thrusts before we felt comfortable with the controls. Almost at once, the mood in the cave completely changed. The nervous anticipation around whether the ROV would even work was replaced by a playful excitement; what could this thing actually do?

The lights from the robot lit up the water, creating a vivid display of the interior of the underwater cavern. The lighting caused the cave walls to radiate deep blues and purples. With precise control, the robot descended into the depths. Watching it dive made my heart flutter. I couldn't believe we had come this far. After all the designing, testing, and re-designing, it was really starting to become clear: we had accomplished an amazing feat of collaborative creation. Building this robot was a product of collective passion and commitment. We had to overcome a myriad of design and technical challenges to arrive at this point. We set out to make a capable underwater ROV that could be used for exploration, using only off-the-shelf parts and tools that are accessible to everyone. Also, we wanted it to be far cheaper than the commercial products that were available. And we had done it.

I didn't just take pride in what we built but also in *how* we built it. The design was Eric's baby, something he originally conceived and muscled into the world. But the current version of OpenROV—the model that made the trip to the cave—was a distant relative of Eric's original prototype. This model was something much greater. From our very first conversation, Eric and I decided to make the project open source, meaning we release the designs, production steps, and bill of materials online for anyone to see and use. We created a website, OpenROV.com (*http://www.openrov.com*), where we displayed the build information and also problems that we were encountering. It started out as a way to show our friends what we were up to, but quickly grew from there. A few months into the project, we were getting advice and support from people all over the world, most of whom we had never met, some with extensive underwater robotics experience. The feedback, suggestions, and insights from members of that community were key to overcoming our challenges. By the time we found ourselves in the cave, the project had benefited from hundreds of contributors, spanning dozens of countries.

We ran the robot down the large cavern and into small offshoots that piqued our interest, as shown in Figure 1-3. At one point, the entire group erupted in cheers as we safely threaded the needle of a tight opening in the rocks. We came across a number of interesting artifacts: a long piece of tubing, some sunglasses, and an old lighter. Items you could imagine a group of teenagers dropping in during an afternoon adventure. We spent so long exploring that we eventually ran out of batteries. Luckily, we had navigated the robot back to a point where we could easily fish it out with the attached tether. It was a silly and humorous mistake in an otherwise successful maiden voyage.

Figure 1-3. Searching in the cave (photo by Brian Lam for The New York Times)

We didn't end up finding any treasure in the cave, but it didn't matter. We had built the robot we dreamed of and, more important, had an adventure doing it. We met hundreds of new friends and collaborators and discovered there were a lot of other people interested in what we were doing. The process was far more valuable than the outcome.

For me personally, the real treasure was never gold, but something far more precious. This maiden voyage of our little robot was a tremendous experience, but my journey started long before that day in the cave. My challenges were more fundamental than any technical design. I had gone from non-existent engineering or design experience to making substantial contributions to underwater robotics. A project that had seemed intimidating and impossible to me only a year earlier shaped me into a completely new person. I had flipped the switch from being a passive consumer of life to an engaged, creative participant in it. I had gone from Zero to Maker.

It all started on a June morning—six months before the trip to the cave—in a small office in Los Angeles. That morning unfolded like most others. I was in early before any of my coworkers had arrived and was busy answering emails and responding to client issues. I didn't expect that it would turn into a judgment day of sorts.

As a startup, we were struggling; revenue had trickled to a halt, investors were backing away, and the attitude around the office was bleak. The plan was to meet at 9:00 am for a team meeting and strategy session. When the founders of the company arrived late and asked only me to come into the conference room, I knew it wasn't going to be good news.

They were letting me go.

Just like the headlines I had seen for the past two years—more layoffs, jobs eliminated, and record unemployment—but delivered with a piercing stab. It was no longer happening around me; it was my new reality. The next day, as the shock continued to set in, I took a long walk through the hills of Los Angeles trying to make sense of it all. I couldn't help but think back on the events that led up to this moment, trying to excavate some sign I overlooked in the haze of unshakable confidence in being on the right path: a good college education, strategic work experience, and a job with a promising young startup company. Then suddenly, on a sunny Tuesday morning, it was gone.

I walked for hours and came to the realization that this was bigger than just losing a job. More important, I felt that in this work shake up, my life story had been stripped away from me. My personal narrative—my sense of purpose and direction in the world—no longer made sense. I had spent so much time justifying my actions (and time spent as a slave to a computer monitor) with the rationale that I believed in the mission of our company. I tried to get back on track mentally by telling myself I'd get another job. I dusted off my resumé—something I hadn't needed to do in years—and just stared at it. I re-formatted and updated my experience, but after all the tweaks and sorts, something still wasn't right. I kept questioning myself: What was I doing, really? No matter how I told my story, I realized,

I couldn't hide one glaring fact: The only thing I was qualified to do was to sit in front of a computer.

To make matters worse, my anxiety over being jobless was compounded by a blooming awareness that I was in a completely wrong business to begin with. It so happens that a year before I lost my job, I had attended a Maker Faire based on a friend's recommendation. She thought I'd enjoy the crowd and the eclectic nature of the gathering. She was right. The Faire blew me away. The interesting projects —robotics, crafts, and massive installations—were only outdone by the passion and energy of their creators. In my wildest imagination, I could probably conceive of a few of these contraptions and characters, but never all in one place—in this bizarre environment where giant unicycles and autonomous robots blend into the crowd. Most strikingly, I couldn't believe these individuals and groups were able to actually build this stuff. I didn't quite know how, but I wanted to be more like them. Thinking and learning more about what I'd seen at the Faire that day led me to Eric and his ambitious plan to build his own submarine. I wanted to help with the robot adventure, though I wasn't sure how I could contribute. Without even a basic high-school shop class education, let alone any kind of engineering degree, I had felt disqualified from even trying.

The jobless wandering and the maker longing were a powerful mixture in the days and weeks after being laid off. The more I thought about it, the more I realized how tragically specialized I had become. I was extremely well prepared for a job that no longer existed, without the fundamental skills I could repurpose elsewhere. I seemed to be far away from being able to build, fix, or create anything of tangible value—any real, physical thing. My so-called skills—emails, social media, and blogging—were hollow substitutes. Now, after hurtling in and out of a digital career, I felt as though I were missing a critical piece of my humanity.

Over the course of the following weeks, my awareness of my manual illiteracy only grew. I met a carpenter at a flea market who was selling handcrafted tables and desks. He explained to me that he used his tables to pay the bills while he pursued a comedy career in the evenings. I envied his resilience. His woodworking skills were something no one could take from him. Unlike my startup job, no one could tell him to stop making tables.

Soon, my desire to re-educate myself with basic making skills overshadowed my worry about finding a new job. I found myself thinking that getting another job would just be a distraction to a bigger goal, delaying the inevitable recovery of a missing vital element of my education.

I wanted to do something about it, but I wasn't sure where to start. I decided to begin with the only lead I had: *Make:* magazine. In addition to putting on the Maker Faire, *Make:* publishes a quarterly how-to magazine filled with interesting projects and makers. It also publishes a popular blog at Makezine.com (*http://www.makezine.com*).

I wrote out a long email explaining my situation to the *Make:* editors, highlighting my suddenly free schedule and dedication to learning the skills and tools that I felt I'd missed out on. I proposed that I would do my best to become a do-it-yourself (DIY) industrial designer before my savings ran out, and blog about my entire experience for the *Make:* blog. I packaged the whole idea under the title "Zero to Maker in 30 Days" and I sent it off.

It was a shot in the dark, but lucky for me, they liked the idea. And now I had a written commitment to follow through on, regardless of how it turned out.

What started off as a one-month commitment to learn new skills turned into a life-changing journey. I quickly discovered that my initial trip to Maker Faire was simply the tip of the iceberg. I continued to meet more makers—a growing community of people who have adopted and rewritten the idea of DIY—and they were nothing like I expected.

Before I dove into making, I barely knew which way to hold a hammer. I wasn't sure if I could fit in or how the *Make:* readers would receive my eagerness to participate. All the makers I had met seemed brilliant, whereas I felt like an average guy, genetically disposed to being uncoordinated and uncreative. How was this going to work?

I had some preconceived ideas about makers: who they were, how they worked, and how they learned. I imagined the process to be a long, lonely, and tedious study of engineering, tools, and science—skills I had bypassed on the fast track to be more "marketable." As it turned out, my initial assumptions were completely off. I quickly realized that these preconceptions were, in fact, the toughest obstacle I would need to overcome. When I recognized how unfounded they were, my own inner maker was able to come crawling out of his shell.

First, I learned how makers really worked. My first trip to Maker Faire left me with the impression makers were lone geniuses, toiling away in garages or workshops, putting countless hours into a project, repair, or invention and coming together once a year at Maker Faire to show off their creations. This couldn't have been further from the truth. Making is definitely a team sport.

Makers are, above all, a connected and collaborative bunch. They meet online and share ideas on forums, blogs, and discussion groups. They give away their designs and collaborate on projects with people all over the world—the exact opposite of the competitive secrecy I had come to know in the corporate world. They've pooled resources to create "fab labs" and "makerspaces," which are physical spaces that serve as hubs for sharing costs and maintenance of larger tools and equipment. It didn't take me long to understand that little of anything is being done "yourself." Making is actually not about DIY, but rather all about DIT, or Do-It-Together.

The next realization came when I began to learn about the new tools these makers were using. Before my immersion, I had a sentimental notion that DIY was about bringing back a bygone era, a time before hammers and nails were replaced with video games and iPads. I imagined DIYers to be torch-bearers, keeping alive the methods and craftsmanship that were marginalized by the onslaught of computer screens and advertisements. I wanted making to help me connect with something I felt had been lost over the past few generations—a part of being a self-sufficient human that was missing from my life.

In a way, makers are the guardians of this industrious self-reliance I had hoped for, but together they're so much more. They understand and respect their place in history, as part in a long line of tool makers and tool users. Although they keep traditional knowledge alive, they are also busy inventing and bringing new technologies into the world. And these are not your grandparents' tools.

The new maker tools are byproducts of increasingly affordable computers, components, and sensors. They are fueled by the rapid exchange of ideas on the Internet and are empowering individuals and small groups with a whole slew of new personal fabrication tools. Laser cutters, 3D printers, and other computer numerical control (CNC) machines (automated machine tools) are now affordable enough to be purchased for a home or office workshop and capable enough to create customizable, consumer-ready products. A product that 15 years ago cost hundreds of thousands of dollars to prototype and produce can now be created with a downloadable file and access to one of the numerous makerspaces that are popping up in cities all over the world.

And learning to use these new tools was shockingly easy, as I discovered. When I started, I predicted I would need an industrial design or mechanical engineering degree before I could make anything useful or valuable. I never imagined I could come so far in such a short period of time. In only a few months, I was 3D printing, teaching others how to use the laser cutter, and designing basic parts in computer-aided design (CAD) programs. I had started welding, working with sheet metal, and creating plastic molds. I was clearly not a master welder and certainly not the best microcontroller programmer, but I knew enough to get started. Anything I didn't know—how to use a machine, what material to use, how to assemble something—I could just pick up on the fly. I learned skills as I needed them, depending on the specific problem facing me. And I was never alone. All the makers I met seemed to specialize in one area or another, and everyone was happy to teach what they knew. In fact, I discovered that everyone still had a lot to learn, but we were all able to leverage one another's skills and knowledge.

As soon as I let go of my misconceptions, I was welcomed into a community of possibility. I realized I was part of something larger: a maker movement. I also discovered my experiences were not unique. This is how all of the new makers were informally inducted. In exploring this new world, I saw a new side of myself, a part that revels in the process of creating and sharing with others. I learned what I was capable of, and it was far more than I imagined.

Even though these observations of "Doing-It-Together" and learning from one another were a revelation for me, I soon discovered that this radical collaboration had been there from the start of this new maker renaissance—rooted all the way back to an experimental class at Massachusetts Institute of Technology (MIT) over a decade earlier.

In 1998, MIT Professor Neil Gershenfeld and his colleagues dreamed up a "fab lab," an assembly of high-tech machines that could build other machines, which he describes as using "supersonic jets of water, or powerful lasers, or microscopic beams of atoms to make—well, almost anything."[2]

The biggest problem they encountered was that none of the students knew how to operate the new tools, so they decided to teach a semester-long course that would serve as an introduction to the fab lab. Thus, the class "How to Make (Almost) Anything" was born.

The class was originally designed as a primer for a small group of advanced students, but quickly evolved into something more as a hundred—from nearly every academic discipline—tried to enroll. The class was a huge hit and was taught for many subsequent semesters. The experience gave Gershenfeld a glimpse into the future of personal fabrication and much of what he saw surprised him, especially with regard to how students were learning. In his book *FAB: The Coming Revolution on Your Desktop—From Personal Computers to Personal Fabrication*, Gershenfeld describes the scene in his class:

---

2. Neil Gershenfeld, *FAB: The Coming Revolution on Your Desktop—From Personal Computers to Personal Fabrication* (Basic Books, 2007).

The final surprise was how these students learned to do what they did: the class turned out to be something of an intellectual pyramid scheme. Just as a typical working engineer would not have the design and manufacturing skills to personally produce one of these projects, no single curriculum or teacher could ever cover the needs of such a heterogeneous group of people and machines. Instead, the learning process was driven by the demand for, rather than the supply of, knowledge. Once students mastered a new capability, such as waterjet cutting or microcontroller programming, they had a near-evangelical interest in showing others how to use it. As students needed new skills for their projects they would learn them from their peers and then in turn pass them on... This process can be thought of as a "just-in-time" educational model, teaching on demand, rather than the more traditional "just-in-case" model that covers a curriculum fixed in advance in the hopes that it will include something that will later be useful.

The "just-in-time" learning model that Gershenfeld described jumped off the page. It was exactly the way I had learned about making. And it wasn't a coincidence: this is how all makers learn.

In my first entry on the Zero to Maker column, I mentioned that my goal was learning enough to be dangerous. At the time, I had no idea what I was getting myself into. I made the comment because I wanted to set the bar low enough that I could achieve it. I didn't expect to become a master of any of the tools, trades, or technologies. Instead, I just wanted to feel them with my own hands and learn how they worked. I wanted to see what was possible.

This turns out to be the best possible strategy I could have taken. After talking to other makers, seeing how everyone operated, and reading second-hand accounts like Gershenfeld's *FAB*, I realized that was what everyone was doing: exploring what is possible.

In retrospect, it seems silly that I was ever nervous about getting started. There was only one lesson I needed to learn. Actually, it was a choice. I had to choose to become a beginner, to get comfortable with mistakes, to ask a lot of questions, and to seek out the right teachers. After I crossed that bridge, everything else fell into place. Makers are a community of beginners, and we're all learning together.

It's easy for me to say, without hesitation, that my quest to become a maker changed my life. But more than that, it has become my way of life. The quest to re-skill myself turned into a fundamental re-thinking of how I view opportunity. And I'm not alone.

What started as a series of garage inventions and side projects has turned into a budding industry, with makers of all different shapes and sizes turning their fervor, skills, and ingenuity into businesses and careers—turning their passion and creativity into entirely new business models based on community and collaboration instead of the old model of cutthroat competition.

The businesses take many different forms. Some are a throwback to traditional craftsmen; artisans that create largely custom and specific pieces of work for a small community of clients and customers. People like Joel Bukiewicz, a knife maker in Brooklyn, discovered that there is substantial demand for his handcrafted cooking knives. After struggling for many years to find work as a writer and suffering a crisis over his career direction, Joel turned his attention toward making and quickly fell in love with the process of creating knives. But his story isn't a harrowing tale of a spurned writer succumbing to isolation and madness a la Stephen King. Instead, in Brooklyn Joel discovered a vibrant community of other makers who share and collaborate to support one another's businesses.

When I asked Joel about his business, he couldn't stop talking about how valuable this environment has been to his development. As soon as he opened up a physical store and showroom, his business took off. He was learning from his audience: what they liked, where and how they were using his knives, and how much they would pay. It was more than a store; it was a catalyst for building his community.

Internet platforms like Kickstarter and Etsy combined with new creative communities like the one Joel discovered in Brooklyn have created a new economic infrastructure for these 21st century artisans to prosper.

Small, community-oriented artisans are not the sole constituency of the movement. Makers are also the driving force behind the proliferation of technologies and platforms like 3D printers, CNC machines, and microcontrollers. Fast growing companies like MakerBot Industries are building and selling desktop 3D printers based on open-source designs.

When I was just getting started with making, I kept hearing about 3D printing. Everyone was talking about it and I had no clue what it meant. It was originally described to me as something very similar to a regular inkjet printer, except that instead of putting ink onto paper, it lays down a thin layer of plastic. Layer after plastic layer, it repeats the process until it has created an actual 3D object. The process continued to baffle me until I actually sat down with a MakerBot and learned how to use it. At its core, the process is as easy as clicking print and waiting twenty minutes for your creation to appear inside the machine. Watching the MakerBot in action helped me to understand what all the commotion was about; there's something magical about printing out an actual, tangible object from a set of digital instructions.

The global community of hobbyists-turned-entrepreneurs has taken 3D printing, a technology that once was only available to researchers and wealthy corporations, and made it affordable enough to be purchased by an individual or small group and used in homes and offices in addition to academic or corporate research facilities. Instead of supporting proprietary research and development arms, these new 3D printing companies have innovated by openly sharing their designs and allowing their communities to give feedback to the product development. Drawing from the open-source software playbook, this model of open-source hardware is enabling small actors and teams to compete with much larger corporations and established businesses because of its leaner and more flexible approach, a strategy I'll cover extensively in Chapter 5. MakerBot and the others have a ways to go before their affordable desktop printers are as capable as the expensive, proprietary models, but they're doing an excellent job of making them easy for new makers, like me, to get into the game. And the maker tools are getting cheaper, more capable, and easier to use every day.

Large corporations are watching this trend, too, and making big bets that this new form of distributive, small-batch manufacturing takes hold. Corporations like Autodesk are busy building design software that enables new makers to quickly pick up the CAD skills they need to get started designing parts and components. Companies like Ford are becoming major partners in makerspaces like TechShop in order to give their employees access to cutting edge equipment. They are betting that innovation comes from empowered, front-line employees. By encouraging their employees to tinker with projects they're passionate about, the companies are hoping to unlock creativity that previously had gone unrealized. Suddenly, making is relevant for more than just the tinkerers and hobbyists who do it for fun. It's a new skill set that can help employees advance in larger organizations.

These major trends—tech-enabled individuals and community-based business models—are all pointing in the same direction: opportunity. In a time when job and career uncertainty are at an all-time high, it's refreshing to see a budding industry (many industries, actually) with so much potential. The maker movement is waiting for people like you to figure out what's next. To use a skiing metaphor, the mountain is covered with a thick blanket of fresh snow—you can go in nearly any direction, but you have to carve your own path.

This book is meant to be a map. It's meant to give you a view of the maker landscape and get you up to speed as efficiently as possible. I have made the transition from Zero to Maker myself in just a few months and witnessed countless others do the same. Based on those lessons, I have created an easy-to-follow formula for avoiding the pitfalls and hurdles that can hold you back. This book is meant to put you in a position to make anything you want, even (and especially) your own business. It is designed to enable. To use the skiing metaphor again, think of this book as the chair lift—carrying you over the freshly covered slopes to give some perspective and dropping you off in a position to get started on your own thrilling run.

DIT (Do-It-Together)

I thought I was pretty original: starting from the very beginning, getting back to the basics, and really trying to understand how things are made. But about three months into my Zero to Maker journey, I came across a story that made my approach and experience seem pretty tame. I learned about Thomas Thwaites and his heroic attempt to build a toaster from scratch. He started with the rawest of materials—copper, iron ore, melted plastic—and set out to end up with the cute little appliance that graces many a kitchen counter.

His story began in 2008 when Thwaites, then a student at the Royal College of Art and Design in the United Kingdom, first hatched his now infamous Toaster Project.[1] His inspiration was a line of science fiction from Douglas Adams' *Mostly Harmless*, one of the *Hitchhiker's Guide to the Galaxy* installments:

> *Left to his own devices he couldn't build a toaster. He could just about make a sandwich and that was it.*

The presumption here is that we are looking at a bemused human being on a distant planet, with outsized expectations to civilize low-tech species that inhabited the world. The hero, however, quickly realizes that without the support of the entire human species, he cannot muster the technological know-how to accomplish the feat of creating a toaster.

This single line piqued Thwaites' curiosity. Was Adams' conjecture right? Have we drifted so far away from the things we use that we are completely unable to recreate the simple objects that are ubiquitous in our everyday life? Thwaites set out to test his theory by building an appliance of his own. Not just a device that toasted bread, which could be done by building some type of a fire burning oven, but to fundamentally recreate the $3.99 toaster (the cheapest model) in his consumer appliance catalog.

---

[1]. This site (*http://www.thetoasterproject.org/page2.htm*) has links to the book, videos, and writings throughout the project.

I couldn't get enough of Thwaites' story; I read through the blog, watched the videos he posted online, and read his book several times. On a primal level, it seemed as if we were scratching the same itch: a lack of control, or input, over the objects and technology that make up the world around us. We were both coming to grips with our manual illiteracy, the disappearing fix-it mentality of our grandparents' generation. But whereas I was dwelling in the sadness of my own missing knowledge, Thwaites was highlighting a larger, more systemic point. The toaster seemed to have been a perfect challenge: an everyday object most people use regularly without a single thought given to its inherent ingenuity and utility.

For Thwaites, the first step in the process was acquiring the toaster he had in mind and breaking it down for parts. He needed to understand exactly what he was trying to re-create. As soon as he started to dig in, more questions arose:

> ...157 parts, but these parts are made of sub-parts, which are themselves made of sub-sub-parts. Does the variable resistor that controls the toasting time count as a single part? But it's made of eight sub-parts, so perhaps it should count as eight? Does a capacitor count as one part or eight?

After completely disassembling and laying out the nearly 400 components built from roughly 100 different materials, he quickly realized the enormity of his endeavor.[2] I knew exactly how he felt. I had run into a similar quandary. With the vague goal of re-skilling myself, I quickly ran up against the enormity of my quest: What did I actually want to make? What tool or tool family should I start with? Should I practice my woodworking skills or spend time learning about 3D printing?

Thwaites wisely opted to redefine the scope of his project, and decided he would recreate just the main operating system of the toaster, or in Thwaites' words, "the bare minimum from which I think I can make a toaster that retains the essence of *toasterness*. These are: steel, mica, copper, plastic, and nickel."

Even after he scaled down his goal to recreating only 5 of the nearly 100 materials, Thwaites still had an enormous challenge on his hands. He had no idea where to get the materials, or even where to start looking. His initial, simple question had evolved into an exposition of how unthinkably hard it is to make anything, let alone do it by yourself.

---

2. *The Toaster Project: Or a Heroic Attempt to Build a Simple Electric Appliance from Scratch*, by Thomas Thwaites (Princeton Architectural Press, 2011, page 18).

Following bizarre leads and random tips, he traveled around the entire UK visiting abandoned mines and digging up the actual, raw materials. At one point, Thwaites attempted to actually smelt iron ore in his microwave (something I strongly advise you don't try at home). With every twist and turn of his adventure, Thwaites' Toaster Project emphasized the unfathomable lengths to which an individual must go in order to find and use these fundamental building materials. By the end of the extreme experiment, Thwaites created something that very vaguely —in both shape and function—represented the original toaster and, according to Thwaites, apparently worked for a brief moment before the 240 volts pouring through unprotected copper wires annihilated the fledgling device.

The project was a huge success in proving the complex interdependency of our world. Thwaites discovered that the novelty of Do-It-Yourself, or DIY, is misunderstood; or as he phrased it, "The point at which it stopped being possible for us to make the things that surround us is long past." Thwaites' toaster insights were a big revelation for me. His project epitomized much of what I had learned over the past few months but had struggled to articulate. Through my own determination to try to do things myself, I actually found a deeper appreciation for how much we rely on one another. I realized that making anything, especially the complex tools and machines that we use daily, requires a dense web of collaboration. When I was able to accept that fact, my anxiety dissipated.

Making is about sharing ideas, tools, and processes. The most prolific makers I met weren't the people who did everything themselves. In fact, they were the individuals most skilled at navigating the web of collaboration and adapting it to their will.

Initially, the concept of DIY created a mental image of a lone inventor toiling away in his basement workshop or a MacGyver-type know-it-all. It was precisely that stereotype that kept me away from making for so long. I didn't have an engineering degree. I didn't know how to use most of the tools in a workshop, and calling me "uncreative" would be an understatement. I figured making was something I just didn't get—it was for *them* and not me. But making, as I discovered early on, was about the art of finding other people—seeking out teachers, creating and joining like-minded groups, collaborating with strangers—and co-creating together. As long as you have an initiative to get started, it quickly evolves to Do-It-Together, or DIT.

It is difficult for a new maker to fathom how it all fits together and even more difficult to see how to contribute to the process. However, after a few basics are out of the way, such as learning the maker lingo, finding the right people, and getting access to the appropriate tools, the making quickly follows. The curiosity will lead the way.

Speaking a New Language

"Excuse me... I'm sorry, you're doing what?" I asked. Surely I didn't hear the gentleman correctly.

"We're working on technology to apply 3D printing to home building." He replied. He went on to describe his business, in which they're working to use cutting edge technology to create actual, life-size homes from a printer. Just like your inkjet printer at your home and office, except much larger and capable of printing in three dimensions. He added an important caveat by explaining that the project and technology were still in the theoretical stage at this point, but my mind had been sufficiently blown.

This was the first I had heard of 3D printing. More strikingly, perhaps, this was the first conversation I had at the very first Maker Faire I attended. That day, and all the sights and experiences that comprised it, will be etched into my mind forever.

I had heard about Maker Faire from a number of different people. Each one had raved about the event. I would ask, "Maker Faire? What's Maker? It's an event, like a conference?"

Their responses were always somewhere along the lines of "Oh, you just have to go." Apparently, it was something to be experienced and not described. As soon as I saw the advertisements for the May event, I made sure to block it off on my calendar.

When the big weekend finally rolled around, I was pretty excited to see what the buzz was all about. I convinced my friend Peter, who was visiting from Germany, to join me. We decided to take the Caltrain, about a twenty minute ride from San Francisco to the Faire grounds in San Mateo. On the train ride down, we shared what we knew about the event and tried to predict what we would experience.

Despite our elaborate imaginings and discussion, the event wasn't anything I had expected, in size or impact. The gates were covered with colorful banners. The crowd was amazingly diverse: families of all shapes and sizes and ages, some dressed up in full costume. Despite the drastic demographic differences, all the attendees, from families with kids in strollers to groups of young adults, wore the same expression of jovial curiosity about what they might encounter that day. Peter and I got our tickets and proceeded through the grounds, still trying to take everything in. Finally, after wandering into the middle, surrounded by makers on all sides, we turned to each other and thought the same thing: where are we? We looked at the program of speakers: DIY Bio, Sugru, Solar Suitcases, Howtoons. So much of it—the people, the sights, and now the language—was foreign to me. I had no clue where to start. We stopped at the information booth just inside the gate to try to orient ourselves. We asked the woman behind the desk what we should see. She smiled with the same look my friends had given me when they told me about the Faire—it was all worth seeing.

Instead of pressing further we resolved to just wander around aimlessly. It was the wandering that brought me into the conversation about 3D printing. I could see why my friends had such a difficult time explaining the Faire to me. Even though I understood very little of what I was looking at or hearing about, I felt strangely at home. Regardless of how little I knew about microcontrollers or machining or steampunk, it was impossible to ignore the passion that each maker brought to his or her project. Their passion bred curiosity and a desire to learn more, and the enthusiasm that was created was infectious. I remember thinking how you don't get to see this kind of effusive creativity first-hand very often. I've certainly never seen it in any kind of office environment or among a large congregation of similarly minded people.

That first day at Maker Faire was the starting point I was seeking. That initial spark of interest would turn out to have an enormous impact on my life. It changed the way I now read and learn, the way I approach problems, and my perception of what's possible.

Two years after the initial encounter I looked up that original program of speakers and exhibitors at Maker Faire. The same words—3D printing, microcontrollers, servo-control—that had been so mystifying to me then were now part of my vocabulary. The tools once completely foreign were now everyday objects in my life. And the speakers and presenters were all familiar names, some of them friends I see on a regular basis. In hindsight, I realize the acquired lingo accounts for a large part of confidence I have to call myself a maker. It took me a long time and a lot of uncomfortable moments to figure it out, but it didn't have to.

How to Speak Maker

Learning the maker lingo is as important to understanding the maker culture as speaking Spanish is to understanding Mexico. It's fine to read a few blogs or just go to Maker Faire, but if you really want to start making things, the first step is to learn the vernacular. Fortunately, the Maker vocabulary is much easier to pick up than Spanish. The tricky aspect of the Maker language is that many of the words are common English that have been repurposed and re-imagined by the maker community. For example, the word "make" is one of the most common words in the English language, but that doesn't mean explaining Maker Faire to a first-timer will be any easier. The word takes on a whole new meaning. "3D" and "printing" are both everyday words with their own associated mental images, which don't adequately describe the new technology used in maker circles. Becoming aware of the new terms is the essential first step.

LESSON 1: GET FAMILIAR

Here are a few terms to get you started—some of the basics. As you'll notice, many of these words are already familiar, but have been adopted by makers with slightly altered meanings.

Making
  Creating and exploring new possibilities through building and experimenting with tools, technology, and materials.

Maker
  Someone who makes or produces something; a person with a propensity to tinker with and create the world around him.

Hack
  A modification to software or hardware; an effective (but oftentimes inelegant) adjustment that solves a problem or fulfills a need.

Kit
  A set of pre-packaged parts needed to assemble a product. Maker kits typically involve electronics and robotics.

Arduino
  A single-board microcontroller. The popular open-source board was created for ease of use and has become a preferred embedded system for maker electronics and robotics projects.

LESSON 2: ASK QUESTIONS!

This one took me way too long to learn. When I began making, I was scared to ask questions. I thought I would sound dumb for not knowing. Once I started, though, I realized that everyone was happy to explain until I understood. Gradually, I gained more and more confidence in asking questions. Now I don't hesitate for a second.

Unsurprisingly, I noticed this characteristic among all the veteran makers I've met: they have absolutely no shame about asking questions, however obvious the answer might seem—whether it's how something works, what a word means, or a request to repeat an instruction. When in doubt, ask!

LESSON 3: REPEAT LESSON 2

Over and over and over again.

Surrounding Yourself with the Right People

You become the average of the 5 people you spend the most time with.

— JIM ROHN

I love that quote. I don't remember where I first heard it, and I'm not sure from whom. I do remember, however, it being a complete revelation. I thought back to jobs and phases of my life, back to college and high school. In each case, no matter how much I had studied or how hard I'd tried, the skills or habits that really took hold could be directly attributed to the people I was spending time with.

I have since adopted this as a philosophy for how to learn new skills. For me, relying on my own initiative doesn't work; I need a foolproof system that keeps me on track to my goals.

Learning the lingo will only take you so far. To use the Spanish language metaphor again, it might be possible to learn the language with only a translation dictionary, but it sure wouldn't be easy. A better way to learn is to practice speaking with a native speaker or, better yet, go for an immersive trip to a Spanish-speaking country. Making is no different. Thankfully, though, you don't have to cross any borders to find an immersive maker experience.

When I first committed to getting started, I thought hard about who I should try to spend my time with, and more important, how I was going to convince them to let me hang around. Not surprisingly, many of the groups and people I sought were people I met (or learned about) that first day at Maker Faire.

I approached that day without specific expectations or knowledge about the weekend, and almost no information about the hundreds of exhibitors, performers, and panel discussions. But because it was our first visit, we wanted to make sure we saw at least one presentation. We stumbled in just as Eri Gentry, who would eventually become one of my maker heroes, took the stage to talk about BioCurious. The tagline for the talk was "the Bay Area biology collaborative lab space." Eri walked us through slides explaining the core elements of DIY Bio. It is a grassroots movement of scientists and amateurs who are working to create cheaper tools and provide access to anyone interested in learning more or experimenting with science, and biology in particular. She provided background information about the high barriers to entry for research, the expense of equipment, and the need for a university affiliation. Then she took us on a photographic tour of her lab—an actual makeshift laboratory set up in the Silicon Valley garage that she shared with four other DIY Bio enthusiasts. She showed us pictures of fume hoods and other lab equipment she and her collaborators had created mostly from off-the-shelf parts. I had no idea what most of the equipment was, but after she put up the slides comparing the costs of commercial equipment to their homebuilt counterparts, it was easy to see the value they had unlocked. She went on to show videos of one of her roommates running experiments on potential cancer treatments in their garage. When Eri shared that her roommate's project had received venture funding and moved into an actual laboratory to continue their research, I knew they must have done something right.

Her thorough, inspiring presentation couldn't have provided a better introduction to the subject and how low-cost, off-the-shelf tools had the potential to support serious scientific research. It placed the maker movement into a new context for me. Then Eri said something at the end of her talk that resonated even deeper: she mentioned she had studied economics in college, which is exactly what I had studied. It took me by surprise because I had presumed my education was actually a hindrance for making. As she left the stage to make way for the next speaker, I walked over and waited in line behind a few other audience members with unanswered questions. When Eri finally turned her attention my way, I was steaming with curiosity. I introduced myself and stepped right in, "You mentioned studying economics in college, how did you learn all this biology stuff? Did you get multiple degrees?"

"Nope, everything I've learned about science I've taught myself. Most of it I've had to learn as I go," she replied, referencing her BioCurious cohort.

"Wow, really? Do you think that's something I could do, too? I mean, starting from knowing basically nothing?"

"You bet! We've got a pretty wide diversity of people who come to our meetings. You should come. Send me an email."

And so I did. I ended up talking with Eri again a few weeks later to get the whole story. She, like me, initially explored the maker movement with only a genuine interest to learn, which in her case was about science and biology. After moving to Silicon Valley and not being able to find an opportunity for an untrained scientist, she set to work to change that. She met a few people who shared her vision of making science accessible and created BioCurious. Eri and her friends began meeting periodically for discussions and experiments. The group steadily grew to more than 500 local members, and pretty soon she found herself near the center of a growing movement, DIY Bio, and was being invited to speak at events like Maker Faire.

Eri's story is a shining example of going from Zero to Maker in short order. Propelled by an eagerness to learn, all it takes is a little organizational skill to bring more makers into your life.

BioCurious is a very small sampling of maker groups. There are dozens of maker subgroups in the Bay Area exploring everything from garage robotics to microcontroller programmers, letterpress enthusiasts to digital fashion. Similarly diverse groups are springing up around the country (we'll go through more strategies for catalyzing a maker community in your area in Chapter 3). Curiosity is a great place to start. If something interests you, the best way to learn more is finding out where and when a group convenes. Social media is also a great starting point. Google+, Facebook, and Twitter provide easy ways to follow makers and makerspaces and stay on top of upcoming events.

Joining a Local Group

Another group I learned about on my first Maker Faire visit was Make:SF, a monthly Meetup (*http://www.meetup.com*) group that offers, in their own words, "an opportunity to get started in the maker community. You can meet local makers, learn some new skills, and grow from there."

Perfect for my derivative goal of spending more time with makers, I knew I'd have to check out the Make:SF group. I looked them up online, and lucky for me, they were having one of their monthly meetings within a few days and only a few blocks from my apartment. The topic of the evening was "Basic Electronics: soldering and assembling basic kits." I arrived at the event a half hour early to try to get a lay of the land. I was nervous. Not only had I never soldered before (which was a little embarrassing) but I had no idea how they would react to my lack of skill. It is one thing to go to Maker Faire and learn about what someone is working on, but it's quite another to make something yourself.

The event was at Noisebridge, a local hackerspace in the Mission neighborhood of San Francisco. A hackerspace, as I'd come to learn, can mean a lot of different things. To some, the word "hacker" can conjure up thoughts of criminally inclined computer experts, working their way in and out of security loopholes on the Internet. In maker circles, however, a hack is something that is modified, either by adding or re-assembling new parts or functions. Basically, it is an alteration that serves a specific purpose. An example would be the Pringles Cantenna (*http://blog.makezine.com/2011/07/05/pringles-can-antenna-turns-10/*), a makeshift device created by Adam Flaherty and friends to extend the WiFi range of their antenna. After a $10 modification to an empty Pringles can, they achieved the same effectiveness as a $150 commercial antenna. Hacks can be done both with software and physical objects. A lot of making is hacking. A hackerspace is a collective or community space that hackers (and makers) maintain to collaborate and share resources. Noisebridge is just such a place.

Outside the building I saw a small piece of tape next to the buzzer that had "Noisebridge" scribbled on it. I pushed the button and waited. I wasn't sure if my early arrival would be welcome, or if anyone was there yet. I took a deep breath to calm my nerves. I slowly opened the door, peeking my head in as I entered. The room was rich with activity, recalling an over-supplied, and under-supervised adult playpen—tools, parts, and machines strung apart everywhere. Most contraptions looked like they were under construction, perhaps so indefinitely. Boxes and shelves lined the walls, filled with more things that could be possibly used for something else someday.

My worries about walking in and interrupting a group of working makers couldn't have been more off-target. If anything, my entrance among a few people scattered around the space—working at computers, combing through the boxes of parts—was completely unobtrusive. They looked to be about my age, young to middle-aged adults, everyone in casual attire. It was a long way from my imagined scene of exaggerated costumes and autonomous robots patrolling the room. I expected the exotic flair of Maker Faire, and wound up in a space that seemed entirely approachable. I almost blended right in to the scene. They probably wouldn't have noticed me at all if I hadn't been standing in the doorway, looking like a deer in headlights.

"I'm here for the Make:SF meetup," I hesitantly announced. "This is Noisebridge, right?"

"Hmmm... I'm not sure. I don't know anything about it." One of them replied.

As I looked around confused, another person came hustling around the corner from the back of the space. He was carrying a box of soldering irons. He introduced himself as Malcolm, one of the organizers of the group. After a few pleasantries, I nervously revealed to Malcolm that I was a total beginner, and he smiled and told me I had come to the right place.

Pretty soon, the room began to fill with other attendees. By the time the session started, there were more than twenty people in the space. Despite the size of the group, Malcolm still took the time to go around and have everyone introduce themselves and say what inspired them to come. The group was strikingly diverse, with men and women from every background: an art director, an animator, an artist, a real estate broker, and a software engineer. When Andrew (the other organizer) asked how many of us were new to making and Make:SF, over half of the hands in the room went up. Clearly this was a safe place to make mistakes and ask questions.

Malcolm divided us into three groups to work on the evening's projects: electronics kits. I broke off into the group that was making the MintyBoost, a device that charges iPhones and iPods with AA batteries. We each received a kit, which in this case was ready-to-assemble electronic parts and a soldering iron. This was my first experience with soldering, and that was obvious. It took me a while to get the hang of holding the iron, let alone melting the solder onto the circuit board. Even though I had only a vague idea of what I was doing, Malcolm walked us all through it. Anything I missed or didn't understand, one of the other group members would step in and help me (and vice versa). It didn't matter that we had just learned it ourselves; if you figured something out before your neighbors, you took the time to help them when they got stuck. We were learning together. By the end of the night, I left with a new iPhone charger made from AA batteries and an old Altoids tin, some basic soldering skills, and a handful of new friends.

The Make:SF experience was the first of many positive group encounters in my Zero to Maker journey. Each one was just as inviting as that first night at Noisebridge. In all my experiences with different maker groups, there's a common thread that underlies them all—a welcoming culture of possibility, encouragement, and collaboration.

Meeting More Makers

Once I opened my eyes to it, I realized that makers and maker groups were everywhere around me. I think you'll be surprised how close you are to the action, too. Here are a few things to do:

1. GO TO A MAKER FAIRE

The maker Mecca. For new makers, Maker Faire is a great place to start. First, the provocative projects and the dedicated people who make them provide the ultimate inspiration for ideas. Second, the diversity of the presenters and exhibits gives you an infinite range of making possibilities. It's a great way to figure out what piques your interest, whether it's with robotics, crafting, 3D printing, or Kinetic Pastry Science Mobile Muffins (which are, in their own words, "delicious, electric-powered, built from scratch, highly maneuverable and capable of 18mph+!"). And lastly, a curious and positive attitude can take you a long way. I attended my first Maker Faire having never assembled anything other than Ikea furniture, and now, only a year later, Maker Faire feels like a family reunion. In addition to the original Maker Faire in San Mateo every May, there are now annual Maker Faires in Detroit and Kansas City, and a World Maker Faire in New York City every September. Plus, there are over a hundred Mini Maker Faires that take place annually throughout the world. It's becoming easier than ever to connect with the larger Make: community.

2. EXPLORE MEETUP

Meetup.com (*http://www.meetup.com*) is a website that hosts a network of local interest groups. Meetups are a great way to meet people interested in...well, basically anything. Maker meetups are no different. I found my way to Make:SF because they were local, but it's very likely there are groups near you. If there aren't any in your area, you can always start your own. Andrew, the original Make:SF organizer, started the group after moving to the Bay Area from New York and finding no groups like the Make:NYC group (now dormant) with which he'd been involved. Make:SF now has more than 1,800 members and has hosted almost 80 events.

3. VISIT A MAKERSPACE OR HACKERSPACE

Maker- and hackerspaces, like Noisebridge, are excellent places to meet other makers. I continually hear and read about new hackerspaces opening up all the time. You can look at a list of nearly every hackerspace on the planet (*http://www.hacker spaces.org*) or check out the Maker Map (*http://themakermap.com*), as shown in Figure 2-1. Just remember that each hackerspace is unique. For example, Noisebridge (where the Make:SF Meetup was held) has a different setup than a place like TechShop. Noisebridge is a co-op model, which works well for experienced makers who need a space to hack, whereas TechShop, which works more like a gym membership, is better suited to makers who need access to tools as well as classes and project mentoring. There's a flavor of hackerspace for every maker type and need (more on this in Chapter 4). I suggest exploring as many as possible to get a sense of what's available.

Figure 2-1. The Maker Map from http://themakermap.com

4. EXPLORE INSTRUCTABLES.COM

Spending actual, physical time with makers is an ideal situation, but it's not always practical, given geographic and time constraints. The good news is that there are some incredible online resources that make virtual time a good alternative. Instructables (*http://www.instructables.com*) is an online community and database of DIY project how-tos for just about anything, from pumpkin JELL-O to an electric canoe. If you can imagine it, there's a good chance there's an Instructables entry on how to do it yourself. If there isn't, you can add it! The great part about Instructables is the community that supports and contributes to it. It's a great way to get feedback for the first-tries or prototypes that you create.

5. VOLUNTEER!

The maker community is a welcoming bunch of people, and it's amazing how much you can learn when you simply offer to help. There are a number of projects and groups that need a helping hand, even from those of us with limited technical backgrounds. You can peruse the Make: site (*http://blog.makezine.com*) or Instructables (*http://www.instructables.com*) for projects that catch your eye and email the maker to see if there is any way you can get involved. Attending a Maker Faire or Meetup event is another great avenue to offer your participation. In Chapter 3, I'll discuss some of my volunteering experiences, which turned out to be some of the most valuable parts of my journey.

Starting Your Own Group

As much as I enjoyed spending time with the BioCurious group and learning from the Make:SF group, I was still missing one crucial element every true maker possessed and I longed for: passion. Eri's exuberance as she described the DIY Bio projects or Malcolm's radiant expression as he showed a first-timer how to use a soldering iron is something that can't be forced or contrived. There is nothing like the inspiration that you emanate when you are pursuing a project you love, something that really gets your heart beating. It's always that kind of maker behind the best projects at Maker Faire. And I was still searching for it.

It wasn't until I met Eric Stackpole that I was able to identify a project I could really pour my heart into. After sharing with a friend that I was looking for a way to combine my interest in the ocean and my desire to start making, he suggested I meet Eric. He had heard that Eric was building a submarine in his garage and assumed that kind of project would align with my interests pretty well. The idea of building your own submarine sounded intriguing enough. If nothing else, I had to hear the story for myself.

Eric and I exchanged a few emails, and we eventually found a time to meet to talk about the submarine, or ROV (Remote Operated Vehicle) as he referred to it. He was quick to explain that it was more of an underwater robot than a submarine. We met a few weeks later on a Sunday morning at a coffee shop in San Francisco. Before we sat down, Eric stopped and asked me, "I have the robot in the car, should we go grab it?"

"Yes, of course!" Even though it was the reason we were meeting, I was a little surprised to be jumping in so quickly.

When Eric pulled the robot out of his trunk I marveled at how small it was—about the size of a microwave oven. And it seemed so refined. The parts looked much more polished and precise than I had expected. I had originally imagined a submarine like the one on The Beatles album cover—a cartoony contraption built to withstand ocean depths with a periscope poking through the surface. Eric's creation was nothing like that. We spent a little more than half an hour going over the technical aspects of the robot. He told me about the tools he used to create some of the parts, and where others could be purchased as off-the-shelf components. He then explained his vision for telerobotics, or creating low-cost, accessible machines that allow people to view and interact with places they wouldn't normally be able to—in this case, underwater environments. "In our email exchange, did I tell you the story behind why I was building the ROV?" He asked.

"No, what's the story?" I replied, interested in hearing the background.

He then launched into his now-famous retelling of the Legend of the Hall City Cave. By the time Eric finished telling me the story, my jaw was on the floor. I was hooked.

That initial meeting was the start of a much bigger conversation. Over the next few months I researched ROVs extensively. I found that they could basically be split into two categories: commercial and homebuilt. The commercial products were clearly well-built and capable tools. They varied significantly in size and capability, but they all had one description in common: expensive. The homebuilt ROVs were also ingenious, but for a very different reason. The creativity poured into making DIY ROVs bristled with energy and ambition. With limited budgets and unshakeable resourcefulness, DIY ROV builders were trying novel techniques and ideas that commercial projects were ignoring. The big problem with the homebuilt ROVs was the lack of standardized and coordinated innovation, not a lack of technology. With a little more organization, we thought, DIY ROVs could be equally effective tools for science and exploration as their commercial counterparts, at a fraction of the cost.

Eric and I started OpenROV (*http://www.openrov.com*), a website and forum dedicated to telerobotic underwater exploration, to organize our discussion, share the original design plans, and invite other collaborators. We made the project open source, which means we release all of the technical designs, specifications, materials, and assembly instructions. Instead of keeping that information proprietary and secret, we're sharing our efforts so that others will be able to contribute to the evolution of the design and participate in the adventure. Aside from just the technical gains we've made from being open source, we've had the opportunity to meet a wide array of interesting and enthusiastic people. We've also discovered, contrary to popular belief, that we can build a profitable business with this open model (which we'll discuss in greater detail in Chapter 7).

Starting Your Own Online Group

It's easier than ever to create a website like OpenROV. The magic lies in attracting a community of co-creators. If there's something you're passionate about making and a group doesn't already exist, consider starting your own conversation. Don't worry about not being finished or having all the answers, just starting the discussion is a big step in the right direction.

The first step is finding out what other groups are in your vicinity or range of interest. Figure out who's working on similar projects and overlapping ideas. I found the best way to do this is to *build a knowledge map*. Before we started

OpenROV, I sat down with a blank sheet of paper and spent several hours googling "homebuilt ROVs," "DIY Submarines," "Open Source Underwater Robots," and so on. Most of the searches turned up similar results and I was able to identify and understand who else was working in this world. I drew a map of how those groups interrelated and overlapped, their relative sizes, and how active they were.

There are two possible outcomes from that exercise: you find a group that's working on your idea or you don't. If you find an active group, perfect! There's a great place to start learning. If you don't find an existing forum that suits your needs, the second step is to start your own online discussion. There are a number of easy-to-use and inexpensive web platforms for hosting your forum, including Wordpress (*http://www.wordpress.com*), Instructables (*http://www.instructables.com*), Google Groups (*http://groups.google.com*), and Ning (*http://www.ning.com*).

After you choose a platform and get a basic site together, the third and most important step is to begin inviting and encouraging the community. The best way to foster an enthusiastic and engaged group is to focus on asking great questions. Chronicle and document your process and progress and invite anyone to participate. Don't try to hide the struggles and hurdles—emphasize them! Makers aren't looking for a finished product; it's the process that attracts them to a project or group. Here are some examples of groups and the platforms they used:

PROTEI—OPEN-SOURCE SAILING DRONE

Platform
 Google Groups

Website
 http://www.protei.org

Story
 Like the rest of us, Cesar Herada and his friends watched helplessly as millions of gallons of crude oil spilled into the Gulf of Mexico after the Deep Horizon explosion in 2010. Unlike most of us, Cesar decided to do something about it. He started a discussion on an online forum, Opensailing.net (*http://www.open sailing.net*), about his concern for the lack of oil spill clean-up technology and his desire to create an autonomous sailing drone that could drag an oil boom. The conversation drew the interest of a number of people on the forum and prototypes of the idea took form. After a series of experiments proved the idea might work, the conversation grew. Soon after, Cesar created his own site for Protei and began to document the progress of the prototypes as well as the steps

they were taking. He used the site to coordinate events and meetups, share designs and updates, and solicit feedback. By creating such a strong community, the group was able to eventually raise over $33,000 on Kickstarter (*http://www.kickstarter.com*), a website on which a project can raise money via a number of micro-donations. The group continues to work on the ever-evolving sailing drone.

DIY BOOK SCANNER

Platform
 Instructables

Website
 http://www.diybookscanner.org

Story
 As a graduate student, Daniel Reetz had a stroke of insight while searching Amazon for neuroscience textbooks. In addition to offering the textbooks, Amazon also suggested Daniel might be interested in a digital camera, which was available for far less than the $400 textbooks. Instead of buying the books, Daniel decided he would try to build his own book scanner. After dumpster-diving for parts to build his 2-camera book scanner, he came up with something that worked. He posted a 79-step instruction set to Instructables (*http://www.instructables.com*), which erupted in a whirlwind of interest and enthusiasm. The community has since migrated to its own website and forum, now with over 1,500 contributors participating in the evolution of the design.

DIY DRONES

Platform
 Ning

Website
 http://www.diydrones.com

Story
 Chris Anderson, then the Editor-in-Chief of *Wired* magazine, wanted to experiment. He wondered if he could embed electronics to make his remote-controlled plane autonomous—to give it an autopilot. His initial experiment began with his kids and a set of LEGO Mindstorms in his backyard, but as the kids lost interest Chris kept working. He posted his progress to the GeekDad

(*http://www.wired.com/geekdad/*) blog, and quickly found kindred spirits. The process inspired him to create DIY Drones (*http://www.diydrones.com*), a community of other autonomous drone enthusiasts that has grown to more than 30,000 members in less than 5 years. The group has collaborated to create a number of different autonomous flying devices, including the ArduPlane and the ArduCopter autopilots. In addition, Chris has spun out a company, 3D Robotics, that manufactures boards and components (largely to the growing DIY Drones community) and is now selling thousands of these devices every month.

WINDOWFARMS

Platform
    Wordpress

Website
    http://www.windowfarms.org

Story
    Britta Riley, an artist and entrepreneur living in New York City, was inspired by a Michael Pollen article about the benefits, both cognitive and environmental, of growing your own food. She wanted to try it herself, but a Manhattan apartment isn't exactly a fertile food-growing environment, especially in the winter. This didn't stop Britta. From her previous work, she had some experience and understanding of how NASA was using a system of hydroponics for growing food in space, by running a high-quality, liquid soil over a plant's root systems. She figured her apartment couldn't be more hostile than outer space, so she began to experiment. She created a blog on Wordpress, Windowfarms (*http://www.windowfarms.org*), and began sharing her design and progress with anyone who was interested, calling and treating them all as co-developers. Windowfarms now has over 18,000 co-developers on its site.

Getting Expert Advice

Thinking about the people with whom you spend your time is more than just a new maker tip. It goes beyond encouragement. In many cases, joining or starting a group is a comfortable way to ease into the process, but it is not necessarily helpful in addressing a specific challenge or roadblock. In some cases, where the technical questions demand a precise, definitive opinion, the most efficient solution is to approach an expert for help. For example, although he was a new maker, Thomas

Thwaites didn't need to join a group like Make:SF or attend a Maker Faire. He had a goal, a specific outcome he wanted to see for his toaster project. Attending group meetings, as friendly as that endeavor would have been, wouldn't have been useful in his quest to recreate the kitchen appliance because he had such unique and specific needs. However, Thwaites still benefited tremendously from the idea of surrounding himself with the right people by getting advice from an expert.

As Thwaites sat before the disassembled toaster, he pondered how his original simple question of whether he can build his own toaster had evolved into a serious engineering challenge. For example, he had to answer questions like "does the capacitor count as one part or eight?" and "what's that white stuff inside the resistor?" Thwaites knew that it was too much to tackle by himself. He had to re-evaluate his objectives. Most important, he realized he needed some advice.

He decided to reach out to Professor Jan Cilliers, Chair in Mineral Processing at the Royal School of Mines at Imperial College in London. Thwaites wrote him an email to see if he had any advice for his fledgling toaster project:

```
From:  Thomas Thwaites <*****@thomasthwaites.com>
To: Jan Cilliers <******@imperial.ac.uk>
Date: 7 November 2008 02:08
Subject: The Toaster Project?

Dear Professor Cilliers,

I'm a 2nd year postgraduate design student at the Royal College of
Art (just across the Royal Albert Hall from your office at Imperial
College). Sorry for contacting you just "out of the blue," but I'm
trying to build an electric toaster from raw materials and I'm in
need of some advice.

As a first step I think I need to get an idea of whether the project
is hopelessly ambitious, or just ambitious. I was wondering if I
could perhaps come to the Royal School of Mines and briefly discuss
the shape of the project?

Yours Sincerely,
Thomas
```

And the response...

```
From:  Jan Cilliers <******@imperial.ac.uk>
To: Thomas Thwaites ******@thomasthwaites.com
Date: 7 November 2008 07:16
Subject: Re: The Toaster Project?

Thomas,This is utterly fabulous! Come see me whenever you can, I
```

    would be happy to help in whatever way I can. Call me on
    ********* first, or email.

    Jan

It didn't take much. Thwaites and Professor Cilliers ended up meeting for lunch that same day. Professor Cilliers peppered Thwaites with questions, starting out with the obvious "why a toaster?" but quickly diverting into the technical aspects like finding and extracting the raw materials. Of course, the conversation with Professor Cilliers exposed some serious challenges to completing the project. That discussion helped Thwaites define exactly what he could do—it helped shape the project. Not to mention, Thwaites now had a friendly resource he could refer back to, which he did many times.

When I read the copy of Thwaites' email and discussion with Professor Cilliers, I did a double take. It looked exactly like an email that I had written to an ROV expert about our OpenROV project. As we began telling more people about the robot, a number of people would bring up Marine Advanced Technology Education (MATE), which is a nation-wide competition for small teams of high school and college students to create their own homebuilt ROV and test them in front of their peers and a panel of industry judges. The goal of the program and event is to inspire more young people to become interested in robotics, specifically underwater robotics.

Eric had competed as a high school student, so he was very familiar with and enthusiastic about the competition. His eyes would always light up when it would come up in conversation. After doing some research on the competition, I found it to be a treasure trove of useful information. It was rich with resources for building ROVs and, more important, it listed the people behind the program, who were ROV experts with a clear interest in getting more people involved with the technology.

One of the initial drivers of the MATE competition was Drew Michel, an ROV industry veteran. The competition was born from a conversation Drew had with Jill Zande, who started the MATE Center (and still runs the competition and organization today). Since they started the program, MATE has held numerous competitions, both regional and national, that have inspired hundreds of students, like Eric, to pursue careers in science and technology. By all accounts, it's an incredibly positive program for kids and school curriculum all around the country. As I read more about Drew, I became increasingly impressed and excited by his resume and list of accomplishments. Aside from helping initiate the MATE competition, he is an internationally recognized author who's won publishing awards for his books on underwater robotics and electronics, a senior member of the Institute of

Electrical and Electronic Engineers (IEEE), and a fellow in the Marine Technology Society. His resume went on and on. The more I read, the more I thought he would be an incredible resource for our OpenROV project—helping us define what was possible, but also give us ideas for some of our technical challenges.

I spent about 45 minutes crafting a thoughtful (and brief) message about what we were building, why we were building it, and how far along we were. It took me a long time to craft such a short email, but the results were worth it. Less than two hours later, I received an excited response from Drew, and we set up a time that he, Eric, and I could get on the phone.

When we all joined on the call, Drew jumped right to some of the most pressing design issues facing the OpenROV: tethering systems, communication, and thrust-to-drag ratios. Drew helped us identify problems we didn't even know we had. After we finished the conference, Eric called me back right away. We were both really excited to have gotten so much great advice. He told me he never would have thought of sending Drew an email.

It really isn't magic. Most people wouldn't think to email someone like Drew, and that's probably the reason experts like him are usually happy to share their knowledge. If you have a specific piece of information that's holding you back or you don't know where to start, chances are there's someone out there with the advice you need. All it takes is a little Internet research, a thoughtful email, and a little bit of audacity.

The benefits of reaching out are greater than just getting advice. It's a great way to get the word out about your project. In addition to knowledge, many times the experts have ideas on other people and organizations to connect with. They can also be a great motivating force. If you know you have the support and interest from someone you respect, you're more likely to follow through. When I talked to Thwaites, he was very emphatic about the value of his relationship with Professor Cilliers for the Toaster Project, telling me that "it was as motivational as it was informational."

HOW TO: Email an Expert

The opportunity to get advice from experts isn't limited to Master's degree candidates with an interesting thesis project. With a little research, it's relatively easy to get advice from some of the smartest people in the world. Before you start firing off emails to Richard Branson and Bill Gates, I'd encourage you to think about two things: *who* and *what*.

WHO

At any given time, no matter what you're doing, there is probably someone else in the world who has done it before and knows it better than you. At the very least, this person knows the silly mistakes to avoid. But how could you possibly know who that is? And how can you expect that they will respond? I like to think about it in three ways:

Avoid the obvious

As we thought about experts who could potentially help us with our underwater robot, people would bring up the famous underwater explorers like Dr. Sylvia Earle or James Cameron. Although it would have been great to get their advice and input, they weren't the right people for us to talk to. Not only is someone like Dr. Earle incredibly hard to reach, but even if we did, the odds of her being able to take the time to provide useful advice were very small. For us, it made much more sense to approach someone like Drew. He was much more accessible and had enough flexibility in his schedule to have a few hours to chat with us.

Proximity helps

If you can, find an expert nearest you, both geographically and relationship-wise. For Thwaites, Professor Cilliers' office was right across the river from the Royal School of Art. For us, Eric had participated in the MATE competition and had received one of the scholarships that Drew established. In both cases, there was a connection. Also, in Thwaites' case, it's much easier to meet for coffee and a follow-up conversation (which they did) when the expert is only a few blocks away.

Do your homework

When I talked to Thwaites about why he chose Professor Cilliers as his mentor, he told me that he'd researched a lot of different people in and around the geology and mining world. What stuck out about Professor Cilliers is that he had the broadest experience. He had a history of working on other outside-the-box projects and Thwaites thought his creative Toaster Project might pique his interest.

WHAT

Thwaites' email to Professor Cilliers and my email to Drew were strikingly similar, and it wasn't a coincidence. There's a simple format for an out-of-the-blue email that tends to generate the best results:

Introduction and basic explanation (lines 1–2)

The email needs some context. It doesn't need to be your life story, but it should give a little background on why you decided to reach out.

Need and ask (lines 3–4)

Be specific and ask for the advice you need. In the cases of the Toaster Project and OpenROV, the maker needed a reality check. Were our projects too crazy? What were we overlooking?

Thanks and availability (line 5)

Obviously, it's best to close these emails by thanking people for their time. Also, set a range of availability. Setting the availability makes the email seem less open-ended and potentially time-consuming for the expert.

Here's my letter to Drew:

```
From: David Lang [mailto:david@openrov.com]
Sent: Monday, October 24, 2011 3:13 PM
To: Drew Michel
Subject: ROV Advice?

Hi Drew,

I hope you're doing well!

My name is David Lang. I wanted to reach out to you about a project I'm a
part of called OpenROV. I started the project with my friend, Eric
Stackpole, with the goal of creating an open-source, low cost ROV that
could be built with off-the-shelf parts. Another aspect of our project is
that we want the ROV to be a scientifically capable robot, unlike a lot of
the other PVC designs.

We've come a long way and now have a community of about 175 people on our
forum. The prototype is ready and we're innovating very quickly. Eric and I
were reading about you and your work and thought it would be worth reaching
out. It would be an utmost privilege to get your opinion on the project.We
can rearrange our schedules to be available for a phone call if there's a
time that works for you.

Best,
David
```

And here's the response:

```
From: Drew Michel
Date: Mon, Oct 24, 2011 at 3:38 PM
Subject: RE: ROV Advice?
```

To: David Lang <david@openrov.com>

David

The good thing about being in my position (semi-retired after 45 years in the industry) is that I don't need to work very hard to put beans on the table and do have time to visit with bright young minds like you and your colleagues and, hopefully, give you the benefit of my many mistakes.

Also, you have contacted me at a good time. I arrived home (my Houston house) last Tuesday after six straight weeks in Aberdeen, Scotland, Kona, Hawaii, Rio, Brazil and my cabin on Belle River in Louisiana, so I am ready to spend some time at my desk. Most days this week work for me.

Before we get on the phone think about 2 things, cables and connectors will be your biggest maintenance issue. Thrust to mass ratio will be your biggest design issue. Mass includes tether drag.

Looking forward to speaking with you.

Drew Michel
ROV Committee Chair and President Elect
Marine Technology Society

HALFWAY THERE...

Finding the courage and inspiration to go out and meet makers (armed with a commitment to get involved) is both trivially easy and impossibly hard. It means getting away from your comfort zone, out of your routine, and into the unknown. But it's also the most important step. Surrounding yourself with the right people (both virtually and physically) is the most determinant factor in becoming a maker.

In the next chapter, I'm going to highlight the most important lessons from my condensed learning period—the wisdom I wish someone had passed on to me when I got started. But everything still stems from the time I spent with other makers. If you remember nothing else from this book, remember that making is a team sport. It's about showing up and exploring together.

3

The Maker Mentality

You don't make it with your hands. You form it with your hands. You make it with your mind.

— EDGAR TOLSON

Nature or nurture? It wasn't a question I had really considered when I started out. I never thought that making could be a hardwired trait. Certainly, I'd met and known a number of people who seem to have a genetic disposition to this stuff— constantly disassembling things, tweaking everything to try to improve it, always busy building something. Those were the quintessential maker traits, of course. But I was still surprised when a reader applied the nature versus nurture question to making. In a blog entry where I discussed my forays into welding, I described a friend who was trying to weld his own grill. In the comment section, one of the readers wondered about the differences between the "make a grill" and "buy a grill" folks. Could someone actually learn those skills later in life? Or was making something you had to be born with?

The question inspired a number of new questions and comments, and eventually became the topic of an entirely new blog post.[1] People weighed in with strong opinions from both sides of the argument. Some thought it was a natural, inborn quality—makers were born, not made. Others swore that those qualities could be learned, and that the right environment and inspiration could ignite the maker flame in just about anyone. Each perspective had completely valid arguments, many with colorful stories and anecdotal evidence. I vividly remember the feeling as I nervously watched the comments role in over the course of the evening and into the following days. I was hoping that the crowd would settle on nurture and that making wasn't a completely hopeless fantasy for me.

---

[1]. The entirety of the ensuing blog post can be found here (*http://bit.ly/13ODsls*).

One reader, Daniel Harrigan, sided on nurture and commented:

> While I'm sure at some level certain people are more genetically pre-
> disposed to making, it can most definitely be taught and encouraged.
> The biggest problem that seems to dissuade people from making (at
> least in modern western culture) is the collective mentality that we
> ought to consume rather than create. Why create solutions when you
> can purchase them? In public American education especially, shop
> classes and the arts are always extraneous programs and rarely part
> of the core curricula. If people were given more hands-on work and
> shown they can create whatever they imagine, makers might not be
> the minority.

Another reader, Ryan Turner, came to the conclusion that making is something innate:

> I've always found machines and robots of all kinds (Discovery Chan-
> nel's "How It's Made") to be absolutely fascinating. But I can show
> people laser cutters, CNC mills (hell I'll even let people use them),
> autonomous model planes... And for most it is forgotten in moments.
> In what universe is this stuff not awesome?

The comments kept coming, and my worry began to evolve. I became less
concerned about whether I could learn the new skills. After all, I'd already spent
the past month becoming familiar with a variety of new tools and could actually see
the progress. Instead, I was worried that I wasn't learning the right skills. With all
this talk about the making characteristic or making gene, I realized that—regardless
if it was natural or environmental—there was clearly a different "maker mentality."
I was learning the tools and equipment, but I hadn't considered the mental aspect.
For whatever reason, makers see the world differently. If making is something to
be learned, understanding the "maker mentality" is a critically important part of
the process.

This was my epiphany. It completely changed how I approached my Zero to
Maker process. I shifted my focus from trying to learn the tools to trying to learn
the mindset instead.

While immersing myself among makers, there were definitive moments where
my way of thinking diverged from theirs. It was those uncomfortable moments
when I had the most to learn. As an admittedly new maker, everyone was full of
advice. Typically, the feedback was along the lines of "just get started" or "make lots
of mistakes." But for me, those types of comments weren't helpful. The initiative

to get started wasn't the problem. And mistakes were inevitable, whether I wanted to make them or not. The maker mentality was something I had to dig out myself, by comparing and contrasting my thinking from the makers I met. After I realized what I was looking for, I identified a number of characteristics I could adopt and improve on.

Even though I had vaguely identified a "maker mentality," it would still be months until I had a clear picture of what it was. These are lessons I'm still weaving together. The maker mentality is multifaceted: from focusing on learning "enough to be dangerous" to sharing everything you learn, from project-based learning to thinking visually. Some of the lessons were sudden, obvious differences between myself and the makers I was with. Others were subtleties I picked up over time. Sometimes I learned because I asked a great question, other times because I made a big mistake.

Of course, I haven't learned everything yet. In fact, the realization that I've barely scratched the surface is part of the maker mentality, an aspect that makes me excited to keep coming back.

Enough to be Dangerous

Don't let not knowing what you're doing stop you from getting started.

— THE ARDUINO TEAM
Presentation at Open Hardware Summit 2011

Only a few weeks after the "make a grill" versus "buy a grill" debate, I had the opportunity to attend the World Maker Faire in New York City. This was the third Maker Faire I had attended, but it was a very different experience. Instead of being an amazed onlooker, I was there to *participate*: Eric and I brought an early version of our OpenROV to exhibit at the Faire.

I learned so much that weekend. I took so much away from the speakers and presenters, many of whom were experienced makers that I had grown to admire. I was absorbing their talks like a sponge. No longer a passive observer to the maker movement, I was now actively doing my best to follow in their footsteps, and each of the talks overflowed with wisdom that I could apply directly to my experience.

Like the other Maker Faires, I was inspired by the exhibitors. In the moments that I could sneak away from the OpenROV booth, I found myself deep in conversation with other makers—discussing their projects, their process, and any advice they had for a relative beginner like me. For example, the DIY Sous Vide Cooker that the couple in the booth next to us had created (more on these two in

Chapter 7). They broke down all the components of their homebuilt contraption, how they built it and how they programmed the Arduino microcontroller to fine-tune the deep fry temperature. They even let me sample a deep fried egg yolk.

The best learning experience, however, was standing at the OpenROV booth and showing off the robot to onlookers. With only an early prototype on the table, ours wasn't much of an exhibit; it was more of an evolving discussion of underwater robotics. Some Faire-goers stayed for over an hour to talk about the design, pulling up chairs to draw out ideas and answer questions. Many of them stopped by multiple times. We learned much more than we taught.

Of all the great people and interesting projects, one conversation stood out among all the rest. Just as the Faire was winding down for the evening on Saturday, Gareth Branwyn stopped by the OpenROV booth to check in. As I mentioned before, Gareth was the Editorial Director at *Make:* at the time and was the one to whom I had initially pitched the Zero to Maker idea. The entire column was really an outgrowth of that conversation. Gareth was my long-distance mentor. Having met many reluctant makers, he understood where I was coming from, and was always there to provide encouragement and support. Over the past few months of writing the column, Gareth and I had traded numerous emails and talked several times over the phone, but this was the first time we had met in person.

I had learned so much in the past few months, I didn't know where to start. With our underwater robot sitting on the table, the conversation naturally started there. I pointed out all the different features, but after an extensive overview of OpenROV and the different components. He seemed surprised how much I knew, and half-jokingly asked me, "So, looking back, do you think you've gone from Zero to Maker?"

Although I should've been more prepared for it, the question caught me off guard. It was the first time I had really taken stock of everything I'd learned thus far. I certainly didn't feel like I had made any real progress; I could still only see the mountain of things I didn't know. But as my mind went through a montage sequence of how much I'd learned over the past few months, I realized I had come pretty far. I thought back to an illustration I had seen during a presentation that weekend from Nathan Seidle, CEO of an open-source electronics manufacturing company called SparkFun. I answered Gareth, "You know what? I think I'm getting there."

I then went on to explain this graphic from the presentation:

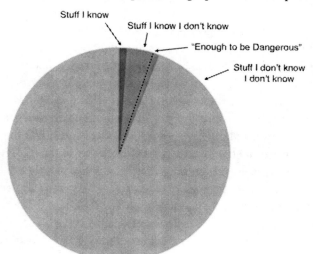

Nathan told us that his goal was to continually expand the "Stuff I Know I Don't Know" slice—that's how he measured his growth. For me, the same is true about my maker journey. Of course, my "Stuff I Know" slice has increased, but not nearly as fast as the "Stuff I Know I Don't Know," and that's great. My original goal was learning "enough to be dangerous," which means knowing how to ask better questions and knowing where to begin looking for answers. Like everything worthwhile, the more you know, the more you realize there is to learn.

I was starting to think like a maker. In my mind, I'd mentally deconstruct everything I came across, wondering how things work, trying to take them apart, figuring out if I could build them myself. I started jumping at the opportunity to fix things; no longer seeing that as a tedious task but an exciting learning opportunity. I viewed the world differently.

Make (and Safety) First, Ask Questions Later

When art critics get together they talk about Form and Structure and Meaning. When artists get together they talk about where you can buy cheap turpentine.

— PICASSO

No maker is an island. Well, except for Tim Anderson. Tim is in a league all his own. National Public Radio (NPR) ran a half-hour feature on him called "Tim

Anderson, Bay Area DIY Superhero." If any one person completely embodies the maker mentality, it's him.[2]

I first learned about Tim online through his ongoing series of updates, "The Free Yacht Saga," on Instructables (*http://www.instructables.com*).[3] At the time (years before my making quest) I was managing a sailing school in Berkeley, California, and was fascinated to learn about seemingly good boats that were being given away. Tim's story in particular seemed pretty incredible—a group of friends who were given an old boat and brought it back to a useful life with salvaged materials and elbow grease. The story was as much about their wacky adventures as it was their DIY techniques. As it turned out, the Free Boat Saga was just the tip of the iceberg.

After reading more of his stories and learning that he was based only a few miles away, I invited Tim to give an evening presentation at our sailing school about what he'd learned during his process. After a series of emails back and forth, we found a time and date that worked. I wasn't really sure what to expect, but after reading through all of the Free Yacht entries I knew it was going to be interesting.

On the night of the event, I was even more anxious and unsure of what was in store. There was a decent turnout for the talk, with the crowd made up of about half sailing club members and half Tim's friends. Tim was running late, and because I hadn't heard from him in days, I began to worry that he'd forgotten. However, as I talked with some of his friends, they assured me that he was on his way. They also had insider's grins that hinted we were in for a surprise. That made me more nervous. He arrived shortly thereafter—without shoes on.

With about 25 people in the room, Tim began the presentation. Listening to him talk about the Free Yacht Saga was much better than reading about it online. Hearing him explain it added an entirely different dimension. The free boat seemed to be symbolic of Tim's worldview: that our culture is full of perfectly useful things that are being, literally, thrown away. He was on a mission to prove that with a little creativity, it's easy to find utility in these discarded items. And besides, building and fixing things is just way more fun.

Each slide had an image of the boat (actually, boats) in progress, accompanied by a hilarious and unbelievable backstory. In one episode, they managed to fit almost 40 people on board for a Fourth of July celebration until the coast guard showed up to spoil the party. On another, he told us the story of when the boat was

---

2. I strongly recommend the NPR feature on Tim (*http://bit.ly/17GZFEs*). The quotes and stories are worth the listen.

3. The Free Yacht Saga (*http://bit.ly/155jHcZ*) is a hilarious and informative read in its own right.

sinking while simultaneously being on fire. I looked around the room at members of the sailing school—more traditional sailing types, most of them older than Tim and friends—and their mouths were agape. This certainly wasn't the safety-first sailing we had been teaching at the school. Many of the members approached me after the presentation and remarked that they seriously questioned the seamanship skills of Tim's crew but, boy, did it look like they were having a lot of fun.

Exactly.

Over a year later, as I arrived back in San Francisco and set my sights on learning to make, I got back in touch with Tim. I emailed him to tell him my plan, and to let him know I was available to help with any projects he was working on. He responded right away, offering suggestions on getting started and inviting me to come work on a new project he was embarking on. He described it simply as "gardening with heavy machinery."

Tim told me to meet him at "the tower," an old air traffic control tower on the decommissioned Air Force base on Alameda island in the San Francisco Bay. The tower had been converted to an office building that housed a number of renewable energy startups, as well as a full machine and fabrication shop. Because it's an old Air Force base, there's plenty of available space to test a kite that can generate high-altitude wind power (which they do), or an inflatable robot (which they've done), or garden with heavy machinery. Tim bills himself as the pro bono night watchman for the startups that inhabit the old facility. As far as I can surmise, this means he keeps a small office and shares tools with the companies.

When I showed up, I asked my way around the building for Tim and ended up finding him in the kitchen, just finishing up a plate of eggs. He saw me and asked, "Oh great, you're here. You ready to build some sustainable infrastructure?"

One of Tim's goals for the year was to grow his own food. For most people, this would involve creating a garden and planting a few vegetables. For Tim, this involved creating massive structures out of reclaimed materials, transplanting fruit trees, and determining how they could survive with only the average rainfall for the area. Definitely not your typical backyard gardener.

This particular day's activities involved turning an old piece of dock from the Emeryville Marina into a raised-bed garden.[4] The old dock—about 15 feet long and 4 feet high—conveniently revealed a four-section trough when it was out of the water and flipped upside down. It was perfectly suited to housing the self-sustaining gardens Tim was envisioning. He walked me through the overall strategy, including

---

4. Tim documented a similar attempt to the Nomadic Garden on Instructables (*http://bit.ly/1d2sy45*).

his design for making sure that these gardens would need as little maintenance as possible. He explained how using hollow caverns created from halved barrels and covered with fast-wicking material would make sure that any precipitation would be absorbed and stored.

After the detailed explanation, I was surprised that the first step was painting the old dock, which seemed to be more about aesthetics than anything else. I actually think Tim started with a paint brush and roller to ease me into it.

After a few hours of painting, the next step was to cut the barrels in half so they would fit inside the hollowed-out dock. Tim handed me a circular saw, gave me a few pointers to make sure I didn't cut my leg off, and let me at it. He wandered away, working on setting up the next task and equipment. I made a few cuts with the saw, playing within the safety guidelines Tim had drawn. By about the fourth barrel, I had it down. I was feeling much more comfortable. By the time Tim came back, I was cutting the barrels like an under-skilled amateur, which was a big step up from a hopeless beginner. Next, we realized that we needed to cut away more of the siding in order to fit the half-barrels in. Tim gave me a reciprocating saw, some general guidelines and, again, wandered off. I had no choice but to try to get comfortable with the tool.

I think Tim knew what he was doing. I think he knew that the best way to start making is to just start doing it. His walking away meant I had no other options. Sometimes, having someone around who knows what they're doing is a crutch, and you can insulate yourself with questions. It's important to get the basics—especially with regards to safety—but after that, the learning takes place with your hands.

Unfortunately, you have to be careful how you use the term "enough to be dangerous." It isn't always the best advice, especially when you're talking about circular saws and other power tools. In those circumstances, the opposite is true: try to learn enough *not* to be dangerous.

That was Tim's teaching strategy—drawing the general safety guidelines and letting me explore. Then I had to build my *own* confidence with the tool.

For new makers, it's both. Learn enough to be dangerous in the philosophical sense, but also enough *not* to be dangerous in terms of safety.

Teach 'em If You Know 'em

"Have you had the forklift lesson yet?" Tim asked me.

"Well yes, but..." I replied with a lot of hesitation. I had, in fact, been shown how to use the forklift, technically at least. My forklift education consisted of a quick run-through of the controls from someone who had also been briefly introduced to the piece of equipment. And that was weeks ago.

"Great! Then you can show Mac how to use it." Tim instructed me, gesturing toward the forklift. Then he walked away, across the expansive lot of the old air traffic control tower, to work on something else. Mac looked at me, bright-eyed and eager for my explanation. I was stuck. I had to explain how to use a very big tool that I'd only used once, and barely so.

I climbed up into the seat of the forklift and looked around, trying to jog my memory by re-enacting the physical motion. I started with the easy things: the ignition, gas pedal, brake, forward, and reverse. After I started identifying the ones I knew, a few more came back to me: opening the gas tank, raising the fork, tilting the fork. Before I knew it, I had almost everything I thought was relevant. Mac jumped in the seat to test out my instructions. As he put the forklift in reverse and started to back up, something didn't seem right. I heard Tim yell from across the yard, "You'll want to pull the lift up before you drive."

"Oh yeah, that too." I echoed to Mac. I should have remembered that.

As soon as I had shown Tim that I was comfortable using a tool, or taking care of one of the tasks, he'd leave it to me to explain it to anyone else who was volunteering that day. Although he never told me explicitly, I think he did that so I would get a better grasp of what I was doing. And I did. It's fairly straightforward to learn something that someone shows you how to do, but it takes another level of understanding to be able to explain it to someone else. With the forklift, I had to quickly think back to my brief experience as well as the explanation I received. When I jogged my memory—going from what was obvious, to what I remembered, to what I suspected—I was exposing the gaps in my understanding. Also, the new forklift driver, Mac, asked me a few questions, which really cemented the points I was uncertain about.

I felt a lot of pressure. Was he going to crash it? Was something going to break? When Tim told us to raise the fork, I was actually relieved. If that was all I missed during my explanation, I had done pretty well. That emotional roller coaster—feeling the pressure of knowing, and the relief of acceptable understanding—is an essential part of building confidence.

Almost every maker I've talked to is eager to share everything they know, and they're ready to help you know it, too. This culture of teaching was in stark contrast to the culture I grew up in. At every level of school, there was only one teacher and that person did all the teaching.

In maker culture, everybody is a teacher as well as a learner. At Make:SF, every meeting starts off with a show and tell of any new projects people are working on. TechShop calls their courses Safety and Basic Use classes, while most of the actual learning happens from experimentation and advice from other members. Every project on Instructables is a step-by-step guide on how it comes together, and nearly every step is full of commenters with tips (and colorful stories) of what might work better, or won't work at all. Every time I've exhibited at Maker Faire, I've spent more time getting good advice and suggestions from others than I did explaining what I've done. It's a big part of maker culture and a critical part of the maker mentality: share what you know.

Where (and How) to Share

The maker world has come a long way since the days of Gershenfeld's class at MIT, but the spirit of the "intellectual pyramid scheme" is alive and well. The propensity to share knowledge—no matter how recently it was acquired, how incomplete the project, or how embarrassing the failure—remains a hallmark of the maker community. Amazing in-person and online communities have formed around the sharing of projects, ideas, and techniques.

Not surprisingly, the list of places to share is the same as places to learn: Instructables (*http://www.instructables.com*), Make: Projects (*http://blog.make zine.com/projects/*), Maker Faire, makerspaces, or meetups. The learning and sharing are intertwined.

For our OpenROV project, we've found that the more we share, the more people share with us, and the better our project becomes. We've worked hard at sharing, and the better we've gotten, the more help we've received. Here are some tips for getting the most out of sharing your work:

Last thing first

Talk about your intention first, not your methodology or plan of attack. Sharing your intended outcome first makes it more likely that you'll receive a novel suggestion—something you might never have thought of. It also helps orient your potential collaborators because it gives them a glimpse of the project from your perspective and puts you at the same starting line.

For example, while explaining the need for a perpendicularly arranged acrylic structure, an onlooking TechShop member suggested we use the strip heater to heat and then bend the acrylic instead of trying to attach two separate pieces. We had never used (or even seen) the lonely strip heater sitting in the corner of the shop, but it turned out to be the perfect tool for the job. Instead of fumbling around with joints and fasteners, we were able to have a better looking and more functional part by using one piece of plastic. (I'll cover the technique in more detail in Chapter 5.)

By first explaining what we needed done instead of what we were doing, we were able to create a better structure in faster time, and at a lower cost.

Document, document, document

Working on a project or learning a new tool takes time and effort. It takes even more time and effort to document your entire progress. Taking photos, writing up steps, and noting all the materials can sometimes take as much time as the project itself.

As tedious as it can be, it's totally worth it.

As soon as we made a stronger effort to document our OpenROV project, participation increased dramatically. Instead of only being a forum for discussion, we posted build diagrams, a bill of materials, and demo videos of how we put the robot together. It made all the difference. The community turned from passive observers to active contributors. We've found that all the extra time we invest in clear explanation comes back to us several fold in the form of good advice and community engagement.

Websites like Instructables make it easy to document your project, too. The projects are presented as step-by-step guides to replicating what you've created. The benefit of presenting your project in steps, as opposed to a finished product, is that contributors can tell you exactly what part of the process to change, refine, or rethink.

Celebrate failure

We all make mistakes, and makers probably make them more than others. The natural inclination is to want to hide or cover up an error.

Talk about your mistakes. If possible, get a video or photo of what went wrong. Not only is it the best way to get feedback, but the photos and videos of the mistakes are often the most entertaining and memorable part of the documentation. People will like you more for your humility, and everyone will remember to avoid that error during their own attempts.

The Persistent-Tinkering Mentality

Hanging around with Tim was a constant look into the mind of a maker. I was continually soaking up information and techniques—taking mental notes and thinking in terms of "OK, now what would Tim do?"

I learned a lot from him. However, successfully mimicking the methods of a maker superhero weren't nearly as enlightening as the big, obvious moments in which I was clearly not thinking like the rest of the group. It was those profoundly uncomfortable experiences when I had the most to learn. One of the most vivid lessons came the first time we water-tested our OpenROV.

Eric and I had finally come to a point in the process at which the next logical step was to test the robot in the water. There were still a lot of question marks, of course, but there was really no excuse for not taking the plunge.

I called my aunt, who lives about 45 minutes south of San Francisco, and asked if we could use her pool to test our robot. Even though she was thoroughly confused, she agreed.

Eric and I met at her house the following Sunday afternoon. Eric brought the robot, which he had taken home after the last time we worked on it. He told me he needed to make some final tweaks before it was ready for submersion. When he knocked on my aunt's door, he had the ROV in hand already connected to his laptop, with the control program up and running. Apparently he was worried it would stop working if he closed his browser. I should have taken that as a bad sign.

We set up our experiment in the backyard around the swimming pool with my curious aunt looking on. We wanted to get video of the experience and decided that Eric would be the videographer, leaving me to control the ROV. As Eric positioned the underwater camera, I tested the motors; they all ran perfectly but were about to be put to the true test below the surface. I lowered the robot into the water, and walked back to the controls. Eric watched intently. I got back to the laptop and fired the forward thrusters (the two back propellers that would push the robot). They worked! Well, they worked for a few minutes anyway, before one of the propellers stopped responding to the signals. We took out the robot and tried again, and it worked again briefly. Pretty soon, though, no matter what tricks or alterations we tried, the ROV was still underpowered.

It was at this moment, when I realized that all of our previous work hadn't accomplished our goal, that I saw the difference between Eric and myself: I was completely useless. There was nothing I could suggest that would be of any help. While I was dwelling on the fact that it wasn't working, Eric was already going through possible solutions, running calculations, trying different configurations, and suggesting alternatives.

My perspective on the moment was different than Eric's. Different in a way that couldn't be attributed to our level of interest or initiative. I was, just like Eric, completely fascinated and intrigued by everything we were working on. I had, just like Eric, the willpower to try something new. Something else was different, though, something deeper than initiative and less formal than a Master's degree in engineering. The difference had everything to do with Eric's incorrigible willingness to try things *another way*. I've come to define that distinctly maker trait as the persistent-tinkering mentality (PTM) and I didn't have it. Not naturally, anyway.

The PTM is tough to describe, but you know it when you see it. It's a combination of unshakeable optimism, unlimited opportunity, and never-ending satisfaction. It's living in a perpetual state of "well, what if we tried..." This was new to me. Eric has a persistent-tinkering mentality. He never sees a project as finished; there's always something that can be tweaked or improved. When a point of potential failure arises, he's already moving down an alternative path.

Up until this point, I had been cruising by these moments of potential failure because I always had someone to lean on: Eric to explain the design, Tim to show me a new tool, an instructor at TechShop to warn me of an overlooked preparation step. But sitting there, watching our robot in the pool, I was in the same place as Eric. We were both looking over the edge of the what-I-know-how-to-do cliff. I was paralyzed by the uncertainty, whereas Eric was busy making a mental hang glider. I asked him about this later, after I had processed my failure for what it was. How was he able to react so resourcefully? He told me it comes from lots of experience, from facing "mistakes" so often that the PTM becomes second nature.

I learned something from that experience, something much more fundamental than how to hold a soldering iron or adjust the settings on a MIG welder. I learned that if I really wanted to become a maker, I was going to have to develop a persistent-tinkering mentality. I have to remind myself every time I face a point of failure, regardless if there's someone more experienced around to lend a guiding hand, that it's an opportunity to exercise the PTM muscle—to practice building those mental hang gliders.

As it turns out, there's quite a bit of science to back up my sinking robot epiphany. Carol Dweck, a psychology professor at Stanford University, is one of the world's leading researchers of motivation. Her book, *Mindset* (Ballantine Books, 2007), describes something very similar to the PTM that I have identified in makers. She calls it a *growth mindset*, which she says is "based on the belief that your basic qualities are things you can cultivate through your efforts."

Dweck compares the growth mindset to what she calls a fixed mindset—a mentality steeped in personal judgement—that situational outcomes are directly related to an individual's natural, unchanging ability. I had exhibited a fixed mindset when the thrusters on our robot failed, attributing the setback to my lack of capability. Eric had exhibited more of a growth mindset. In Dweck's research, people with growth mindsets typically respond to challenges with even more effort. In fact, they never even realize they're failing—they always see it as learning.

During our pool experiment, Eric knew he was learning. I was learning, too, I suppose. I learned that makers exhibit a special type of growth mindset, one that never views a project as complete and is always looking for ways to improve. It sees a sinking robot and thinks, "Hey, at least the electronics compartment stayed dry. We can work with that!"

Luckily, I had Eric there to show me what a PTM looked like. After we pulled the robot out of the pool and assessed what we learned, we brainstormed a list of changes that we could make to improve performance. We were able to think of a half-dozen ideas. I tasked myself with exploring new strategies for waterproofing the motors. This helped me to feel re-empowered, and that our goal was again a realistic possibility. Apparently, according to Dweck, this is exactly what fixed mindsetters should do in these situations. She suggests a way to trick your mind into the growth mindset by making a plan and sticking to it, regardless of self-pity. Easier to say than do, I think, but helpful nonetheless.

I spent the next day exploring new materials and processes. I felt a renewed excitement for the project and was eager to try out the new strategies. Having a check list of new experiments made me reframe the entire experience. I saw it as a valuable learning process. And if the next round of solutions didn't work, at least I was mentally prepared to keep trying.

Cultivating a PTM: Write!

Writing the Zero to Maker column for *Make:* was a very fortunate situation. Through that experience, I was able to meet incredible makers and get advice from all of the *Make:* editors and readers. But more than that, I believe that forcing myself to write that column every week helped me develop a PTM of my own.

Just as Dweck suggests, I wrote down my intention (to learn enough to be dangerous) and plotted out a course to get there. Even when I bumped into hurdles, I knew that I still had to write about it, so I was constantly looking for the silver lining—the learning experience I could write about in the column. It became an unintentional PTM-building tool.

The good news is that you don't need to write a column for *Make:* to replicate this experience. The access to makers is something you already have through Make:Forum and Instructables. And, it's never been easier to create a blog to document your maker progress. You can easily set up a journal of projects or progress online.

It doesn't matter if anybody else reads it. The *process* of writing is the real reward. It's the perfect way to build a PTM. Also, it's fun to look back at earlier posts to see the progress you've made.

You can easily create your own maker blog on sites like Wordpress (*http://word press.com*) or Tumblr (*http://tumblr.com*). This is a great way to do it, especially if you have hopes to eventually create a community or discussion around your projects. You could also email one of the maker communities and ask the members if they have a good place to document progress; basically, the idea is to replicate my Zero to Maker column with a maker community of your own. For example, we'd be happy to have a new maker journal his or her experience with underwater robotics at OpenROV (*http://openrov.com*). In fact, we'd love it. I'm sure many other maker groups would agree.

Project-Based Learning

I had already read about Alex Andon and his incredibly cool Desktop Jellyfish Tank project on several blogs, so I knew most of the easy details: he was a young-ish guy, a marine biologist by training, but his love for jellyfish and his maker spirit had brought him down a different path.[5] Mesmerized by the fluid movement of the jellyfish, he wanted to create a tank for himself. When he learned that they required a specialized aquarium design in order to live in captivity, he—in true maker form —started experimenting with different designs in his garage. His design evolved and his experiments eventually turned into a small custom jellyfish tank design business.

When I learned that Alex was doing all of this from a warehouse mere blocks away from me in San Francisco, I had to stop by and see it for myself.

When I arrived to the address listed on the website, the delivery door was rolled open, exposing the entire office/warehouse to the street. I poked my head in, quite obviously wanting to ask someone a question, and a few folks looked up from their work. It didn't look like a jellyfish operation, but I asked nonetheless. Alex popped up from behind a desk and shook my hand. He was just as affable and excited as he'd seemed in the video. He was obviously passionate about jellyfish; his knowledge of the creatures was obvious as he pointed out details in the breeding tanks.

---

5. The first time I read about Alex and the idea of a DIY jellyfish tank was on the Make: Blog (*http://bit.ly/155mdjc*).

The most interesting part of the tour was hearing Alex describe the process of coming up with his current project, the desktop version of the tank (the one he was selling like crazy on Kickstarter). He told me about his initial prototypes and how he had to solve the simultaneous problems of keeping the water moving without the jellyfish getting sucked into one of the water pumps. It started with modifying existing tanks in his cousin's garage using parts he picked up on Craigslist and evolved from there.

Alex's jellyfish tank is a perfect example of what I've come to call an *Unknown Project*. This is something that no one has done before or a new twist on an old idea. Like Thwaites' Toaster Project or the OpenROV, an Unknown Project is usually outside a maker's comfort zone. It lacks any kind of instruction manual, and sometimes even a clear outcome. With Unknown Projects, the challenge of true problem solving can be both inspiring and engaging as well as highly intimidating. Unknown Projects require in-depth design thinking and usually a lot of failed attempts.

As counterintuitive as it might sound, finding an Unknown Project is actually a great way to get started as a new maker. Picking a really big, seemingly impossible project is a perfect way to frame the experience. It creates a roadmap of things to learn and it also takes a lot of pressure off the success of the final project. Even a failed attempt will be a wonderful story worth retelling. The important thing is the learning along the way. It will open doors you never knew existed. And you never know, you might just stumble into a community of other people who also want to see that idea come into the world.

Once you have an Unknown Project idea, you can start breaking it down into more manageable goals, or what I call *Known Projects* (projects with plans and an expected outcome). That first night at Noisebridge, when I splattered solder around like a kindergartener finger painting, is a perfect example. The event was billed as an "Introduction to Soldering and Electronics." Those were skills I wanted to learn for the OpenROV project, so it put the experience into a larger context. At the beginning of the evening, we had to choose between three different kits, which are a particular type of Known Project. For me, choosing the portable phone charger kit was more relevant than the musical pen, but it really didn't matter. Regardless of which project I chose, the more important aspects of the activity were learning how to manage a soldering iron and practicing attaching resistors to a circuit board. Hell, my phone charger barely worked!

In a class on angular sheet metal that I took at TechShop, we created an aluminum box. There were required tools and materials, as well as directions left by the makers who've come before. These types of Known Projects are great for your first stabs at the soldering iron or creating an aluminum flower box. All these Known Projects have helped me get comfortable with a number of different tools and processes: electronics, welding, laser cutting. And through them all, I inched closer to my goal of building an OpenROV.

KITS—NOT JUST FOR KIDS

Kits are a perfect example of Known Projects. Maker kits come in all shapes and sizes—electronics, robotics, metalworking—and have been used by tinkerers for centuries. They range in complexity from simple LEGO sets to flying quadcopters. Traditionally, kits have been associated with hobbyists such as model builders. However, from a maker perspective, they can be a powerful force for a DIT education.

Michael Schrage, research fellow at MIT's Sloan School Center for Digital Business, wrote an essay for *Make:*'s Ultimate Kit Guide that chronicled the importance of kits in the development of new technologies and industries.[6] He traced their impact back to the beginning of the Industrial Revolution, when James Watt and Matthew Boulton first commercialized their steam engine design by selling kits. Schrage then runs through many of the most important technological revolutions of the past hundred years—automobiles, airplanes, personal computers—and traces their roots back to the hobbyists and kits that laid the foundation.

The importance of kits, as Schrage notes, is a two-way street:

> *Talented amateurs don't just build kits; kits help build talented amateurs. And healthy innovation cultures—and successful innovation economies—need the human capital that their talent embodies. Kits are integral, indispensable, and invaluable ingredients for new value creation.*

A kit is more than just an unassembled product. It's an opportunity to understand the basic workings of a thing. Putting a kit together means going over every step, each detail building the foundation for what comes next. It's also an opportunity to connect with other makers. Popular kits usually have a community of other makers that can share insights, tips, and tricks—a group to ask questions and help get past stumbling blocks.

---

6. The entire essay can now be found online (*http://kits.makezine.com/blog-post/kits-and-revolutions/*). I highly recommend it; it changed how I think about OpenROV and entrepreneurship.

Underwater robots are no exception. It was, in fact, a kit called the Sea Perch that guided many of the underwater robot professionals I met. As I first started to explore the idea, I kept hearing about the Sea Perch. It was inspired by a 1997 book by Harry Bohm and Vickie Jensen, *Build Your Own Underwater Robot and other Wet Projects* (Westcoast Words, 1997). In the book, they share their design for a DIY underwater robot that can be built at home, using mostly PVC piping. The robot was made as an educational tool and was picked up by MIT who, with support from the Office of Naval Research, created a curriculum around it to get more young people interested in ocean engineering. Based on the frequency that people bring up the design in conversation, their strategy must have worked.

The Sea Perch is a perfect combination of low-cost, accessible parts and easy assembly. Classrooms still use it as a teaching tool—the experience exposes students to a number of the challenging aspects of working underwater, especially in regards to buoyancy and propulsion. It's also a point of commonality with other ocean engineers.

Bottom line: seek out kits as part of your maker education.

FINDING THE RIGHT KIT

After you know where to look, it's not hard to find the perfect kit for your maker education. The right kit could be something that helps you work toward your Unknown Project idea or just something you find interesting. Here are a few places to start looking:

Make: Ultimate Kit Guide
    Make: took a lot of the dirty work out of evaluating kits. Their Ultimate Kit Guide reviews over 175 different kits and lays them out in terms of complexity, documentation, and community. This is a perfect place to start.

    An evolving and updated version of the Ultimate Kit Guide (*http://kits.make zine.com/*) can now be found online.

Online communities
    Many of the online maker communities offer kits for new members to get up to speed quickly. Groups like OpenROV, DIY Drones, or Windowfarms all offer kits for a reasonable cost, as well as growing communities of people who are sharing in the experience.

SparkFun/Adafruit/Maker Shed

If your Unknown Project involves electronics, starting with an Arduino Kit from Maker Shed (*http://www.makershed.com*), SparkFun (*http://www.spark fun.com*), or Adafruit (*http://www.adafruit.com*) is a great choice (more on this in Chapter 5).

Embracing Constraints

With making, like everything in life, there are always constraints. Even when you have access to an incredible array of tools, like I did at TechShop, and the knowledge of how to use them, which I was gaining through classes and practice, there are always factors that constrain what you can make. Learning to work in that zone and expecting barriers, both known and unknown, is another hallmark of the maker mentality. During my first maker holidays, I decided I was going to make all my gifts; it was a perfect example of learning to work within those boundaries.

The first constraint I bumped up against was time; I didn't have much of it. With only a week before the holiday, I quickly realized that my utopian idea of creating all my Christmas gifts was far too ambitious. I re-evaluated my plan and decided to lower my goal to just making a gift for my parents, mainly because they were still so confused about what I had been up to during the preceding months. The next combo of constraints was trying to find a gift in the sweet spot of something they would like and be impressed that I made, but also something I could actually make. After racking my brain to think of a gift in the center of that Venn diagram, I settled on creating a cribbage board shaped like the state of Minnesota (where they live). Cribbage is a card game where the score is kept on a pegged board. My parents love playing cribbage, and whenever I visit them it's my favorite thing to do. I can play my mom and dad in cribbage for hours and not get tired of it. We sneak games in before dinner, in the morning while drinking coffee, or block off an entire night and play each other. I knew that the cribbage board would be a gift that looked great and would get a lot of use while highlighting just how much I'd learned.

Now, given that I was familiar with most of the machines in TechShop, one might assume that creating this cribbage board would be a simple, straightforward proposition for me. Unfortunately, it wasn't. It never is. No matter how clear the vision, the road to a completed project is bound to take a few unexpected turns.

The first step for me was deciding what material to use. I thought about making it out of acrylic plastic, because I had become really comfortable with using the laser cutter. This seemed like a good idea, until I checked in with TechShop and learned that the laser cutters were completely booked up for the next three days, which was all the time I had to complete the project. I took that setback as an opportunity to push myself to use the ShopBot, a CNC mill machine, and make the cribbage board out of wood.

The next step was creating the design I wanted to use. I found a shape of Minnesota online and added the necessary peg holes for the cribbage game. The design turned out to be the easy part. The harder part was taking the design and converting it to a cut file in the G-Code language that could be read and used by the machine. Because I was using the ShopBot, I was constrained to making sure my design was something the system could handle—it couldn't have too many small details or sharp turns.

In addition to providing an outline of the design, the software also needs to tell the CNC machine how to cut—how fast to drill, where to start, how deep to go on each pass, etc. To figure out the right "feeds and speeds," as they say in the shop, I needed to decide what size endmill (the cutting tip of the mill machine) I was going to use.

By the time I got to this point, I was pretty far out of my comfort zone. I had only used the ShopBot once before, during the basic use class under the watchful eye of an instructor. For safety's sake, I thought it best to bring in a second opinion. I asked my trusted TechShop Dream Coach, Zack, to review my work so far, and help me decide on the right endmill to use. Zack suggested 1/4 inch for the state outline and 1/16 inch to drill the peg holes. But just as it seemed like everything was coming together, I encountered another hurdle: I had assumed that TechShop kept a stock of endmills that could be purchased (they did), but unfortunately they were out of 1/16 endmills. And because the holes were so specific, we couldn't substitute a different size. Zack said he could order one, but it wouldn't arrive for a few days, and I didn't have any time to spare before my trip back to Minnesota.

As Zack and I discussed the potential solutions—none of which seemed very promising—we were interrupted by someone seated at the table next to us, Martin Horn. Martin and I had met before, frequently sharing projects and ideas whenever our paths crossed at TechShop. He had overheard our discussion about the endmill dilemma and had an idea that might solve the problem. Martin, besides being a part-time instructor at TechShop, is a wizard on the CNC machines. For starters, he helped us clarify the problem: TechShop had 1/16 inch drill bits, they just weren't configured for the ShopBot. Zack and I nodded in agreement. He then suggested how we might modify one of the standard drill bits to work with the ShopBot by using a custom brass rod to hold a standard 1/16 inch drill bit. I wasn't really sure what he meant, so I pressed him for a deeper explanation. Martin, confident in his idea and short on time to explain, replied, "C'mon, I'll show you what I mean."

Martin and I headed into the metal shop to test out his theory. We created it as he explained it: sawing off a portion of some spare brass rod Martin had tucked away (which seemed totally natural at the time, but now I'm kicking myself for not asking why he had it), drilling a 1/16th-inch hole through the center of the rod with the lathe, and using a hand saw to create a slit for compression around the drill bit. Throughout Martin's detour, I couldn't see the big picture of how this was going to work. However, as we went along and Martin's vision began to take shape, I understood his plan. Pretty soon, we had a part that looked like it just might work. It was good enough to try, anyway. I brought the makeshift part back to show Zack. He spun it around in his hand, gave it a thorough look-over, and said the best thing a maker can hear: "You know what? This might just work. Let's give it a shot."

And so we did. I set up the ShopBot with our modified part, and away it drilled. It worked perfectly. After the entire board was cut, I was beaming with excitement. I eagerly ran back to show Zack and Martin. Their responses were similar: slightly happy, interested that our brass rod solution had worked, but mostly not surprised. Unlike the roller coaster of emotion that I'd experience—from excited to be making a gift, to disappointment that I couldn't finish it, back to excitement and accomplishment—Zack and Martin were relatively unimpressed.

They were used to this experience. They had long since learned to embrace the unknowns of a project and relish the making do-ness of an imperfect but workable solution. Martin and Zack weren't worried about the roadblocks. They took them as an opportunity to explore a different path. In fact, they loved the different path. After seeing the smile I couldn't wipe off my face and the finished product in my hand, Martin said, "Pretty cool, right? I love that. It's pretty cool to not only make the thing, but to make the tool you needed to make it."

As I've come to learn, every project is a crooked path. With a maker mentality, the path is the fun part—every constraint is an opportunity to try something different.

Valuing the Process

I was really proud of my cribbage board gift for my parents—probably excessively so. After I made it, I couldn't stop smiling for at least an hour. I carried it around and showed it to anyone one who would take the time to look. It wasn't that difficult to make. In fact, anyone with a basic introduction to CNC machining could create the piece in a few hours of work. That didn't faze me, though. The notion that an experienced machinist could make it with ease is trounced by the pride and fulfill-ment I get from seeing the cribbage board permanently displayed on my parents' living room table.

Even though I was still learning to think like a maker—the persistent-tinkering mentality, openly sharing mistakes, and embracing the winding path—I could def-initely feel the value of what I was doing. It meant a lot to me, regardless of how amateurish my outcomes were.

It turns out, there's actually some science behind the value I perceived I was creating. A group of researchers led by Michael Norton, a professor at the Harvard Business School, published a paper in July of 2011 outlining a phenomenon they dubbed the IKEA Effect.[7] The study sought to explore the relationship between labor and love; and how customers, builders, and makers who have put their own labor into a project value their own efforts. The first experiment used (not surprisingly) IKEA furniture to see whether or not the assembly of the product affected the cus-tomers valuation process. The study participants were split into two groups: build-ers and non-builders. The builders were given sets of IKEA furniture to assemble, whereas the non-builders were given completed sets that they could inspect. Later, they were asked to bid on their furniture. Not surprisingly, the builders tended to bid much more, an average of 63 percent higher.

After their initial test, Norton and his team modified the experiment to include other types of products in an effort to generalize the results. The next iteration used origami frogs and cranes, again separating the group into builders and non-builders. And again, they found the builders valued their own creations five times more than those of others! The team ran another variation of the experiment with LEGO kits, but this time had the builders and non-builders disassemble their kits after completion. They found that once the kits were taken apart, the builders and non-builders valued them equally. In other words, the IKEA Effect had dissipated once the kit was disassembled, even though the builders had valued it higher when it had been fully assembled moments earlier.

I can't say the study surprised me. After spending so much time making projects like the Christmas cribbage board, it was easy for me to recognize my irrational attachment to my creations. The reasoning seemed obvious to me. Using traditional economics, however, I would have a hard time explaining my rationale. The traditional model suggests:

---

7. Here's the full paper on the IKEA Effect (*http://bit.ly/16ZAjRW*).

```
Finished Product Value = Materials/Parts Cost + Labor/Assembly Cost
```

Using that methodology, the IKEA Effect makes no sense: why would people pay more to provide labor? In the conclusion of the study, Norton and the team speculated as to why the builders, or makers in general, created this self-perceived value. Without making a definitive claim, they offer a number of speculative causes for the IKEA Effect. They speculate the feelings of ownership as a plausible cause, citing studies that show people feel ownership over things that are given to them as well as things they've spent time handling. That didn't resonate with me. Again using my cribbage board as an example: I no longer had ownership of it, and never planned to retain it, but that didn't change how valuable I thought it was.

The next series of explanations by the team seemed a little closer to home: the added value was created by the effort. The process of assembling the project and the positive feelings of accomplishment all get baked into a builder's perception of value within the object. This seemed closer to my feelings, but there was still something missing.

For me, it wasn't that difficult; I didn't need a scientific study to confirm something I already knew. The added value of making something wasn't a subconscious, self-projecting bias that I was unaware of. I knew I was getting a lot more out of it. In addition to getting the actual, physical thing, I also got the knowledge of how to put it together, of how it really worked, as well as a story of my making experience. When you buy something at the store or over the Internet, all you get is the thing:

```
Buying = Thing

Making = Thing + Learning + Story
```

When you look at the making process from this perspective, it's easy to see why there's so much more embedded value. For my cribbage board, I had a decent looking product—nothing that would drive a high price on eBay, but something I loved dearly. I gained more experience with the ShopBot and the CNC process. With Martin's help, I learned how to modify the tools I needed to complete the project. And, perhaps most importantly, I have the story about making the cribbage board that my parents and I play on. My parents have a tangible piece of evidence that I wasn't completely wasting my time trying to re-skill myself, and that I actually learned something.

Norton and his team noted that the IKEA Effect dissipated when the project wasn't completed, meaning that if the builder wasn't able to complete their project they didn't attribute a higher valuation. I think that's only part of the story. In my experience, unfinished or unsuccessful projects still hold a lot of value because the learning and story still accompany the process:

```
Buying (Fail) = Crappy Thing

Making (Fail) = Crappy Thing + Learning + Story
```

In fact, sometimes the learning and the story of a failed project become much more valuable than a successful project ever could have been. Take Thwaites' Toaster Project for example. From an ability-to-toast-bread standpoint, the toaster project was a failure. In simple economic terms, the cost of the project could never compete with its store-bought inspiration. In Thwaites' own words, "It took nine months, involved traveling 1,900 miles to some of the most remote places in the United Kingdom, and cost me £1,187.54 ($1,837.36). This is clearly a lot of time, effort, and money expended for just an electric toaster that didn't work... an object that Argos sells for just £3.94 ($6.10)." However, another way of looking at Thwaites' experience was to see him going on a wild adventure through the UK and getting a unique education in manufacturing interdependencies. He ended up with a story good enough to land him a book deal and numerous speaking engagements. And even though it doesn't toast bread, he still has the toaster, which he plans to keep.

> It's certainly something that I'll never throw away, because (to put it cornily) it embodies so many memories... For me, the stuff that really has emotion and meaning attached to it is stuff with a bit of history. The provenance of things is important.

Thinking Visually

For those who harbor serious doubts about their creative abilities, like I did, I think there's another important piece of advice that is sometimes overlooked: start drawing.

It might sound silly or irrelevant, but I think there's something to it. Nearly every maker I've talked to has mentioned drawing as an important part of their process. Some of them have a specific pen or pencil that they love, and I'm no longer surprised to find them carrying a sketchbook for ideas. It was never a direct suggestion or piece of new maker advice, but it always bubbled up in my conversations and interviews with makers. Kent "The Tin Man" White mentioned it during my trip to his workshop in Nevada City, California. AnnMarie Thomas mentioned it in our conversation about her Maker Faire presentation "Making Future Makers."

In his book *Shop Class as Soulcraft*, Matthew Crawford brings up his side interest in drawing (and includes many of his hand-drawn pictures in the book) which, if it weren't for the recurring theme in my conversations, would have seemed otherwise irrelevant to the rest of his meditation on the experience of making things.

I'm not sure I would have noticed the undercurrent of drawing in the string of makers I met if it weren't for the sketching course that I took through a community college a year prior. Before my desire to start making, I had a fear of drawing that I wanted to overcome. A pen in my hand and a blank piece of paper used to send shivers down my spine. Any time I'd try to draw anything, whether it was a map for directions, a diagram for something at work, or just a doodle while I was on the phone, the moment the pen started to run across the paper, I would mentally seize up with doubts about my creative talent. Even my stick figures made me cringe.

The course was a creative godsend. Every Saturday, I'd escape the typical routine of hovering over my laptop to the sanctuary of the Pasadena City College and the creative barrier-breaking activities that had been laid out by our instructor. I loved it. It wasn't a typical art school drawing class because it was completely focused on sketching. I learned techniques to make quick, beautiful, proportional expressions of designs and ideas. The course started right where I needed it to— just letting the pen feel comfortable in my hand. We moved on to lines, then to shading, then to contour. The great part about being such a novice is that you're able to make a lot of progress with just a few simple changes, which was a common theme for all the tools I would experiment with.

I still refer back to the book that guided the course curriculum, *Rapid Viz* (Cengage Learning, 2006), and block off hours of the week to work on my sketching. Not surprisingly, many makers I spoke to acknowledge a similar affinity to that book.

Speaking from personal experience, pushing myself to improve my sketching skills had two important effects:

Being comfortable with The Suck

Inevitably, while you're learning a new tool or skill, there's going to be a period when everything you produce is no good. In fact, it will probably suck. For me, drawing was no different, and to some extent, The Suck continues to this day. I've grown comfortable with it. Learning to muddle through my bad sketches has helped shape my patience for other tools and skills. It's taught me to trust the process.

Visual thinking

When I first started sketching, I was amazed at how horrible I was at judging things in three-dimensional space. As I went through *Rapid Viz* and the course, I learned to think about how the subjects of my drawing would exist in 3D. This process and practice of building these mental 3D models helped me imagine new ideas and how they might work. Being able to sketch the ideas helped me to convey those ideas to others to get feedback and advice.

Be a Fixer

Every morning I drink a kale smoothie. I throw a few leaves of kale, a banana, and coconut water into my Magic Bullet blender. It's a delicious way for me to get more greens in my diet, even though everyone at the OpenROV office makes fun of me for being so "healthy."

My teammates, especially Eric, hate the Magic Bullet. I'm really not sure why. I think it's because, to them, it represents an obnoxious noise and a perpetually dirty dish around the office. As a product, I don't care for it that much, either. It's a fairly cheap, gimmicky, as-seen-on-TV machine that doesn't actually work that well. And after a year of use, the Magic Bullet was really starting to show its age.

Like so many products in this day and age, it seems built to systematically fall apart as soon as it reaches the end of its warranty period (a concept known as "strategic obsolescence"). It is part of a class of products designed for a specific level of crappy, marrying cheap materials with shoddy handiwork. It seemed that they had run tests to find out how long it would be before a consumer would no longer consider the amount of money spent for the product worth arguing about, and then they designed the product to break at that exact moment.

And, we get what we pay for.

Sure enough, about a month outside the "Limited" One-Year Warranty, the Magic Bullet gave a mid-smoothie death growl and came to a screeching halt. It was a Saturday afternoon. I was in the OpenROV workshop by myself, writing this book. Despondent about losing my ability to drink kale smoothies, I wandered back to my computer and immediately brought up the Amazon page to see what a new Magic Bullet would cost. $49.99. And Amazon Prime could have it here by Tuesday.

But before I gave in to my weakest consumer instincts, I figured I'd try to fix it. At the very least, I'd take it apart to see how it really worked.

I picked up the device and twisted it around in my hands, trying to find the angle of attack. I'm not sure if you've ever tried to deconstruct a blender, or more specifically a Magic Bullet, but they don't make it easy for you. So difficult is it, in fact, that I actually paused to wonder about the legality of trying to open it up. It felt like I was trying to break into Fort Knox.

Upon further examination, I finally found my way inside. I unscrewed and pulled apart as many pieces as I could, even though many of them seemed like they would never go back together. I laid them out on the table, step by step, leaving a trail that I hoped would guide me back to re-assembly. I was learning more about how the device worked as I peeled back each layer.

As soon as I had it all dismantled, Eric walked into the office. He looked curiously at me, huddled over the table with an array of screws and parts strewn about. I confessed, "The Bullet stopped working. I'm trying to fix it."

"Dude, just get a new blender." He replied and laughed. He walked over, surveyed my mess, and told me, "A great engineer always knows when it's time to let something go."

"No. I really think I can get this figured out." I replied. And I really did believe that. Now that I had everything apart, I could see how the blender should have been working. I diagnosed the problem as a faulty spring that had corroded and stopped working. It seemed like the bad spring was preventing the motor from activating and the blender from turning on, but there was only one way to find out, so I switched it out.

I started to put the blender back together, steadily following the trail of parts that I had laid out. I wasn't sure if my solution would fix the problem, but I was slightly more confident. Knowing how the blender worked gave me a new perspective on owning it. Somehow it became more valuable.

As I put the final screws in place (slightly surprised by how smooth the reassembly went), I called Eric back over for the moment of truth. I had a grin on my face, "I think I got it."

He smiled, too, probably still doubting me. Then, with collectively held breaths, I pushed the blender down to turn it on. Nothing. I don't know why, but for some reason I was expecting the blender to turn over like a car ignition that hadn't been started in a while, slowly sputtering back to life. But there was nothing. Not even a click. Eric laughed and walked away.

Despondent, I began running through a mental checklist of what else could be wrong. And then I saw it. My fix and rebuild was a success in all ways but one: I forgotten to plug it in.

I plugged it back in, turned it on, and the Bullet let out a confident roar. Eric turned around, still laughing. A rookie mistake.

Fix It First

No matter how much I highlight the ways in which it's getting easier for makers to get started, it still remains a challenge, given the realities of our daily lives. Market forces, like low prices and convenience, have created an arms race for product unfixability. It goes beyond just making it difficult to fix products, with many products actually *prohibiting* it. Admittedly, I had grown accustomed to the throw-away lifestyle. If something broke, oh well. The cost of replacement usually trumped the hassle of repair. It wasn't until my maker journey that I truly recognized the cost of all the cheap (and unfixable) products in my life.

In the back of my Makers Notebook is a Maker Bill of Rights, as seen in Figure 3-1. It is based on the idea that "If you can't open it, you don't own it."

makezine.com

THE MAKER'S BILL OF RIGHTS

■ Meaningful and specific parts lists shall be included.
■ Cases shall be easy to open. ■ Batteries shall be replaceable. ■ Special tools are allowed only for darn good reasons. ■ Profiting by selling expensive special tools is wrong, and not making special tools available is even worse. ■ Torx is OK; tamperproof is rarely OK.
■ Components, not entire subassemblies, shall be replaceable. ■ Consumables, like fuses and filters, shall be easy to access. ■ Circuit boards shall be commented.
■ Power from USB is good; power from proprietary power adapters is bad. ■ Standard connectors shall have pinouts defined. ■ If it snaps shut, it shall snap open. ■ Screws better than glues. ■ Docs and drivers shall have permalinks and shall reside for all perpetuity at archive.org. ■ Ease of repair shall be a design ideal, not an afterthought. ■ Metric or standard, not both.
■ Schematics shall be included.

Make:
technology on your time

Figure 3-1. The Maker Bill of Rights

Here are some tips and resources for developing a "Fix it First" mentality:

The lowest price doesn't mean the lowest cost

Do your homework on products before you buy them. Keep in mind that buying the cheaper product can oftentimes end up costing you more in the long run. After my Magic Bullet experience, I did some searching around the Internet for reviews and instructions on repair, and almost nothing came up (except for similar stories of products failing just outside the one-year warranty period).

After hearing a friend rave about her Vitamix blender, I looked that up, too. Here was the first Amazon Review:

454 of 461 people found the following review helpful

★★★★★ **Vita-Mix is fantastic!** November 4, 2004

By Kate McMurry   TOP 500 REVIEWER   VINE™ VOICE

I have owned the Vita-Mix Super 5000 since 1997, and I love it. I can grind flour, make juice, soup, ice cream (the fruit smoothies I make with this are to die for)--in short, every kind of food processing you can imagine. This baby is built to last and has a 7-year warranty.

In 2004, a couple weeks after the 7 years were up, I was very upset when it broke down. I thought, "Isn't that just the way these things tend to go with warranties?" I searched for Vita-Mix on the web (the URL is their name, and there is a link on this page, to them, too). I clicked on the button for service, got their 800 number, and called them at 6:00 p.m. Eastern time, and they were there. The service rep was friendly and knowledgeable. She asked me to describe in detail the symptoms of my ailing Vita-Mix. She then said, when I asked about it, that even though I was past the window period to do so, the company would still allow me to buy a three-year extended warranty for $120. I jumped at that, and found out that I could use it to cover the parts (and shipping) to fix the problem. She said I needed an "assembly piece" that sold for $75 without a warranty, a "drive socket" worth $12.00 and the shipping of $9.95 would be no charge to me. She said she was including a small wrench and instructions and I can fix it myself. So right off the bat I used most of the $120 to fix my first problem.

I've never seen anything like this kind of service. When our tabletop water distiller broke, we had to send it in for warrantied repair, which cost $50 shipping (we paid), and weeks to get back. I think shipping the Vita-Mix to the company might have cost nearly that much, so I was delighted to have this option of getting the parts in 10 days to fix it myself and have my beloved Vita-Mix back working within a week and a half.

If you are ever in the market for a food processor that is professional quality that works great with amazing customer service, this is it. The price is very competitive for what you get, and the product itself is fabulous. (Only one tiny little caveat--which is true for =any= blender--wear earplugs when you do the very high speeds or you could hurt your ears.)

Update March 23, 2009: I've owned the Vita-Mix 12 years now, and I still love it! See my comment to this review with today's date which contains information on my experiences with the sales personnel of the Vita-Mix company today.

Update February 12, 2013: My Vita-Mix is still going strong--16 years and counting!

16 Comments | Was this review helpful to you?   Yes   No

That's it! That's exactly the type of review you'd hope to see for a product. This type of review isn't always at the top of the Amazon page. Instead of searching for "Product X review," I suggest searching for "How to fix Product X" or "Product X repair." Those search results are usually a lot more maker friendly.

iFixit

When you're taking apart a Magic Bullet, it's very apparent that the manufacturers never intended for you to get inside of it. The complicated and confusing ways it goes together, matched with hidden and difficult-to-access screws create a puzzling process. Not surprisingly, they don't advertise the method for repair.

Gone are the days when (most) products came with repair manuals or spare parts. Luckily, the Internet is filling the void. The site iFixit (*http://www.ifixit.com*) is the central hub for many of these missing manuals. They have thousands of repair guides, ranging from installing a new dock connector onto your iPhone to troubleshooting your Kenmore washing machine. They also have a parts store that offers many of the common tools and materials needed to fix or refurbish an old device. And, their forum provides a way of tapping into the collective knowledge of other industrious members of the iFixit community.

Most of the disassembly guides are well documented, with lists of tools you will need, videos, and pictures for every step. This was the first place I checked for the Magic Bullet instructions. Unfortunately, they don't have that one yet. Maybe I'll create it!

Fix-it clinics

As I've learned for all aspects of making, it always goes better when lots of people are involved. Fixing products and things is no different.

Fix-it clinics or repair cafes are a popular part of many Maker Faires and Mini Maker Faires. I've also started to see them organized as events at makerspaces around the country. If your makerspace has the appropriate tools, organizing a fix-it clinic can be a fun event as well as a great way to get a number of other perspectives on fixes you're working on.

ENOUGH TALK, TIME TO MAKE SOMETHING

With the DIT and maker mentality concepts out of the way, there isn't much keeping you from at least *starting* the project you've been dreaming of, or getting your feet wet with the maker movement (and all the exciting opportunities that it might unlock).

The next questions to address are about tools: *What if I don't have them? Where can I get access? Which tools should I learn first?*

The next few chapters will deal with these obstacles, and I'll show you that you already have access to many of the tools and machines you need to create a prototype of anything.

4

Access to Tools

My preconceived notion of makers wasn't limited to the people I had imagined: the mad-scientist, MacGyver-types. I also had preconceptions about the *places* where making happened. I envisioned elaborate workshops and laboratories; private places such as garages and basements, filled with tools, contraptions, and works-in-progress. All of the profiles I read and makers I had met reinforced that mental image. They all had incredible workshops, each customized to their specific flavor of creation.

I didn't have that type of workshop. In fact, having taken residence on a sailboat in San Francisco Bay, I didn't have much of anything. No workspace or tools, except for a few wrenches and screwdrivers for basic boat maintenance. It didn't take long for me to realize that having access to tools was going to be a critical issue, no matter what maker route I decided to take.

"I Think I'm Going to Need a Bigger Boat!"

In the first conversation about Zero to Maker, during which I explained my blurry vision of overcoming my fear of jumping into the DIY culture and chronicling the journey, Gareth (my soon-to-be editor at *Make:* magazine) mentioned *Make: Electronics* by Charles Platt. It was a book *Make:* had published on basic electronics, and Gareth was particularly proud of how it was structured; it catered directly to the complete beginner. It assumed nothing. Someone could start with no background in electronics or grasp of the terminology and end up with enough information and skill to tinker with electronics projects. I decided to order it because electronics was something I wanted to learn, and I was eager to see what *Make:* had to offer people like me, who are approaching the subject with a blank slate.

The Amazon package with the *Make: Electronics* book arrived a few days later. I was excited and full of determination. Gareth had given my column idea the go-ahead, and one of my first assignments was to write about how a beginner learns electronics. I blocked off a full day to read the book and begin experimenting. I was about five pages in when I discovered a major obstacle: I didn't have any of the required tools. Gareth was right when he said that I didn't need any prior experi-ence, but in order to complete the exercises and hands-on projects, I would need a whole series of items, such as a soldering iron, batteries, and a multimeter. I real-ized that despite my commitment to learning and my eagerness to get started, my goal of going from Zero to Maker would be impossible if I didn't have the right tools.

This hurdle also cast a spotlight on my next challenge: cost. Tools are expensive! Unless you're using them regularly or working on a big project, it's very difficult to get enough use out of a specific tool to justify owning it. For a beginner who only wants to experiment, it's just not reasonable.

I looked over the *Make: Electronics* required tool list, mentally adding up the costs of acquiring them. And these were just the tools I needed to finish the *first* book on the *only* subject I had attempted. There was so much more to learn. Of course, I knew that the tools would be useful for future electronics projects, so I could look at them as an investment, but who knows what other tools I would need. Not to mention, the materials costs were already sunk. I wouldn't be able to get multiple uses out of a dead battery, used solder, or anything I was likely to break in the process. My wandering mind quickly saw the expenses spiraling out of control.

In an effort to keep the Zero to Maker journey as frugal as possible, I committed myself to finding creative ways to avoid spending money at every step of the process. As it turns out, there are a number of new and evolving solutions for people in my shoes. Most of them involve some form of collaborative consumption, or services that provide all the benefits of access, without the hassles of ownership, such as initial purchase costs, taxes, and maintenance. In other words, sharing.

As I would come to learn, sharing tools does more than just defray costs. It has a fortunate by-product of creating community, connection, and a wealth of collective knowledge about the different tools.

The trick with all these new services and communities is knowing how they work and what you can expect. Just as you might imagine, the best solution for tool access depends on what you plan to make. Each service has its limitations, but knowing the capabilities of each is a great way to effectively find what you need, when you need it.

The Ultimate Maker Tool

Before we go any further with the maker tools, I want to share a piece of advice I got from Andrew Sliwinski about the ultimate maker tool: Google.

Andrew is a jack-of-all-trades type of maker. It's tough to pin down exactly what he specializes in, but he can prototype just about anything. I first learned about him in a *New York Times Magazine* profile (*http://nyti.ms/17H2yFe*), which also discussed the hackerspace he started in Detroit (he's currently building a web app for kid makers called DIY—more on this in Chapter 9). I was so intrigued and inspired to hear what these "Kitchen Table Industrialists" were creating, I sought him out to see what advice he had for new makers.

By the time we met at a Mini Maker Faire in Oakland, I was deep into the Zero to Maker process. I described to him the maker mentality and how I wish I had heard it articulated prior to getting started. It would have saved me so much time and eliminated so much unnecessary fear.

He laughed and nodded his head. Based on his experience teaching maker workshops to beginners like me in Detroit, he knew it all too well. It was important, he thought, to develop a specific curriculum to help address that exact need, so before he started his maker workshops—full of welding, electronics, and 3D printing—he would give a short lecture on the importance of utilizing Google in the making process.

He would pull open a browser in front of his students and explain that although it might seem daunting, there is likely an answer or explanation only a web search away. He encouraged the students to utilize that tool, not to get stuck, and, most important, not to try to learn everything. Over the course of the workshop, he was going to give them some basic ideas on getting started, but the full curriculum involved their own imaginations, curiosity, and the world's best search algorithm.

And wherever Google leaves off, YouTube (usually) picks right up. They're the yin and yang of maker knowledge on the Internet. Google provides all the explicit knowledge you could want: measurements, material properties, parts lists, and the like. And YouTube provides just about any form of assimilated knowledge you could ever hope to find: the instructionals, the video tutorials, or what Alexis Madrigal described as "a momentary apprenticeship" (*http://bit.ly/18HBarP*). The *what* and the *how*. The explicit knowledge is the hard facts and details. The assimilated knowledge is the subtle way of doing things, like the grip and swing of a hammer.

The transfer of maker skills involves both types of learning, with each perspective giving the other more context. Used together, Google and YouTube hold the answer to many of your burning questions. Take full advantage.

Hackerspaces, Makerspaces, Fab Labs: What's the Difference?

My initial despondency regarding the lack of tools and the high costs didn't last long. After my Make:SF experience at Noisebridge and the first few classes at Tech-Shop, it quickly became apparent that I *do* have access to most of the tools I need, I just have to know where to look and what to ask for. The hackerspaces, maker-spaces, and fab labs that are emerging are, so far, the best solution for accessing tools.

To be perfectly honest, I still have a hard time with the hackerspace/maker-space/fab lab terminology. Which is which, what location has what tools, how to join and participate; the lines are blurry.

The most important thing to know about these spaces is that, regardless of whether they share the same self-appointed definition, like hackerspace or maker-space, each and every place is unique and different. Even among the chain of corporate-run TechShops across the country, some locations have tools that others don't, some emphasize motorcycle repair, and others have more maker business incubation. Each space reflects the community that emerges. Also, the spaces aren't mutually exclusive. For example, as a maker in San Francisco, just because you're a member of TechShop doesn't mean you can't frequent Noisebridge, and vice versa. Each of the spaces and communities offers a different flavor of maker know-how. It's the fluid overlap between all spaces and groups that makes the local and global maker community so valuable.

Here are some brief overviews and histories of each of the space types and how each of them is best utilized:

Hackerspaces

It's humbling to remember how new this modern maker movement really is. In 2005, there were only a dozen or so of these types of spaces around the world. Their rarity was matched by their novelty. They represented an anarchic model of decentralized creation. John Baichtal recounts a brief history of hack-erspaces in his book, *Hack This*,[1] from an underground collective in Europe, to scattered clandestine groups across the United States, to their current incar-nation of co-op-style creativity hubs.

Now, as I write this in 2012, there are more hackerspaces than I could possibly count. Thousands, definitely, with probably double that amount in

---

[1]. John Baichtal, *Hack This: 24 Incredible Hackerspace Projects from the DIY Movement* (Que Publishing, 2011).

some level of the planning stages. An awe-inspiring map of all the hackerspaces in the world can be found at *http://hackerspaces.org*. You'll be surprised how close you are to the action.

"Hackerspace" seems to be the original term for this type of community space. And, the spaces that consider themselves "hackerspaces" mostly maintain that grassroots, co-op style of organization. As such, many of the tools and machines in hackerspaces are maker products (like a MakerBot Thing-O-Matic or a derivative of the RepRap 3D Printer), donated or secondhand gear. In my experience, hackerspaces also tend to be a place where works-in-progress collect, including experiments that are ongoing, repairs that are being undertaken, and project builds that are coming together. If I wanted to debug an Arduino microcontroller, take apart a Microsoft Kinect, or work on a CNC machine that has fallen into disrepair, I would head straight for Noisebridge (or whatever the nearest hackerspace happened to be).

It's worth noting that there are now subcategories of hackerspaces, like those for kids (more in Chapter 9), those for Burning Man projects, and those for biology. It's all available; your curiosity is (usually) the only price of admission.

Makerspaces

In some respects, makerspaces are one of the emerging subcategories of hackerspaces. All of the self-identifying "makerspaces" are listed in the directory at *http://makerspace.com*. Most of them carry the same hackerspace ethos of community-centered creation and collaboration. However, the makerspace term has evolved to mean something different: a little more professional, more mainstream type of hackerspace. That's just my opinion, and someone could (quite fairly) argue that I'm over-simplifying things, but I believe the distinction can be helpful for new makers.

TechShop is probably the first true makerspace. It's a for-profit business; its model consists of members who pay to take classes and attend events, and it supports a wide array of well-kept tools and equipment. TechShop in San Francisco has over $750,000 in tooling, including numerous Epilog Laser Cutters, CNC vinyl cutters, ShopBot CNC machines, and a Waterjet CNC machine. It's all there. It's going to cost a little more than a co-op-based hackerspace, but you can also expect more consistency with the tooling, instruction, and availability. The makerspace is a place to go if you want to prototype an idea, receive professional instruction, or use the shared tooling to run a small business.

Following the TechShop model, a number of other makerspaces have sprung up around the country and the world. These include places like Makerhaus in Seattle, 3rd Ward in Brooklyn, and the aptly titled Makerspace in San Diego. I expect to see continued excitement and evolution in these types of spaces in the next few years, with newer hybrid models like Artisan's Asylum finding unique ways to blend makerspace quality tooling, hackerspace quality community, and the very real need of actual square-footage space for makers to work on projects.

Fab labs

Just as the hackerspace is the grassroots, co-op take on a maker community and the makerspace is the business-oriented approach, the fab lab can be seen as the academic take on making powerful production tools more accessible. Here's the fab lab concept, originally defined and described by Gershenfeld in *Fab* (Basic Books, 2007):

> As you wish, "fab lab" can mean a lab for fabrication, or simply a fabulous laboratory. Just as a minicomputer combined components—the processor, the tape drive, the keypunch, and so forth—that were originally housed in separate cabinets, a fab lab is a collection of commercially available machines and parts linked by software and processes we developed for making things.

The first fab labs consisted of laser cutters, CNC vinyl cutters, CNC mill machines, and tools for programming microcontrollers. The goal was to create machines that could replicate themselves, "until eventually the labs themselves are self-reproducing."

A grant from the National Science Foundation (NSF) helped get the first fab labs going in a few different locations around the globe, and the project has continued to evolve. In addition to fab labs, the derivative FabLab @School program is aimed at getting digital fabrication tools and curricula into secondary education.

Here a Space, There a Space, Everywhere a Makerspace

To really understand these new creation spaces, you have to get to know the people who inhabit them. Better yet, meet the people who are creating them. Jeff Sturges is one of those people.

From a young age, Jeff seemed to be destined for makerspaces. For most of his life, he wandered through different worlds, following his curiosity into educational and vocational opportunities that could feed his hunger for cross-disciplinary knowledge.

As a kid, Jeff was a good student, but his real love was for tinkering; he was the type to pull things apart and learn how they worked. When other kids were playing with RC cars, he was the RC car mechanic. As he grew older, the tinkering turned to real cars, dirt bikes, and four-wheelers.

He decided to attend Middlebury College, a small liberal arts school in Vermont that both of his parents had attended. When he showed up, he was shocked to learn that the college didn't have an auto shop. He thought to himself, "Wait, what am I going to do here?"

He quickly found a maker respite in computer tinkering and AV club, absorbing any technical education he could find. Not surprisingly, he ended up becoming an IT technician and network administrator after school. But he couldn't quell his curiosity, and eventually decided to go to Cranbrook Academy of Art to study architecture. After that, he worked for a real estate developer, an architecture firm, and a design firm, all while confidently (and unknowingly) chasing his dream of cross-disciplinary creativity.

It wasn't until he stumbled across the hackerspace concept that he realized what he was meant to do. As one of the early members of NYC Resistor, Jeff helped work through the challenges and opportunities of one of the original community-centered hackerspaces. He saw what worked and what didn't. At the same time, he was volunteering at the South Bronx Fab Lab, which gave him an even broader perspective on the creative potential of these spaces. Jeff was the earliest of early adopters.

Taking what he learned from NYC Resistor, Jeff decided to move to Detroit because it was an area in deep need of creative reinvention. He thought it could use a mixing and amplification of existing resources. The city also had an ample supply of affordable and flexible space. It was a perfect storm of opportunity to apply the makerspace model in his own way. And Jeff was the man for the job; he even did it twice.

His first attempt was very similar to the NYC Resistor model—the classic hackerspace model. He sent out an email blast to his friends and fellow makers in the Detroit area and outlined his intentions. His friends responded right away, and planning commenced. Pretty soon, they had scoped out a space and were moving in. OmniCorp, one of Detroit's first hackerspaces, was born.

Jeff wasn't satisfied, though. Setting up OmniCorp was fairly straightforward given his circle of friends, but it wasn't achieving the truly democratizing social impact that Jeff believed was possible with makerspaces. Just as Gershenfeld was bringing Fab Labs into underserved areas and developing countries, Jeff wanted to bring these tools into the neighborhoods and communities in Detroit that could really use them. Which, for a white kid from New York, wasn't exactly an easy sell.

Jeff knew he needed engagement from the community, and found a great partnership with the Church of the Messiah. Pastor Barry Randolph became a great advocate of the project, even offering space in the church basement to make it a reality. Jeff has continued to work directly with the leadership there, and the Mt. Vernon Makerspace has blossomed.

Jeff's experience is proof that the democratizing of tools and technology is possible if we all work together. He demonstrated that makerspaces can succeed anywhere there is a committed community, even in a church basement.

CREATING A MAKERSPACE

Over the past few years, I've seen and visited dozens of these spaces. I've seen some of them fail, but most of them flourish. In every instance, I always make a point to ask organizers and community leaders what makes a successful makerspace. What goes into it? What are the ingredients? What is the special sauce?

I've received a number of insightful responses, but Jeff's description seemed particularly useful. He told me it's a combination of people, space, and tools. Of those three, the people are the most important. Once you have the people, the other two become straightforward.

Fortunately, it's easier than ever to get your own makerspace off the ground. In fact, *Make:* has helped create a 60-page "Makerspace Playbook" (*http://maker space.com/maker-news/makerspace-playbook*) that guides you through everything from choosing a location to designing for creativity to generating project lists. It even includes sample liability waivers. The Makerspace Playbook was designed specifically to encourage schools to build more creative curricula, but it also doubles as a wonderful resource for starting your own makerspace.

Step 1: People

In each case, Jeff organized the people long before they ever had a space to inhabit or tools to use. Sometimes it only takes an email out to a group of friends; other times it means building relationships with community leaders such as church organizers or school officials. If you're starting from scratch in your area, you can employ a lot of the same techniques we mentioned in

Chapter 2, like using Meetup (*http://www.meetup.com*) or finding similar interest groups to help build your local maker posse.

Step 2: Space

After you have a foundation of community support, the next decision is to start scouting out a potential location. If we've learned anything in the past few years, it's that just about any place can be used as a makerspace. There's been a surge in libraries evolving into makerspaces, makerspaces making their way into schools, or just in a neighborhood garage. The space should reflect the needs of the people who will be using it.

Depending on your goals and community, it doesn't have to be a stand-alone space. Like Jeff's experience with the Church of the Messiah, you might be better off tapping into an existing resource like a museum or library.[2]

Step 3: Tools

Flipping through the contents of the Makerspace Playbook, you'll find "The Perfect List" in the Tools and Materials section of the book. Here's what it says:

> *Ha ha! We don't have it! Equipment lists are as individual as the space and its members.*

The truth is, for many spaces, the right tools are the ones that you can get your hands on. Many makerspaces have found success advertising themselves as places that people in the community can donate their old tools to to give them a second life. Others, like Artisan's Asylum in Somerville, Massachusetts, have created a model that allows for members to donate tools and machines in return for a discounted membership fee. Jeff recalled that in the early days of NYC Resistor, the community decided that they really needed a laser cutter, so a number of them chipped in to buy one. Jeff told me that it was one of the best investments of his life, and one of the driving forces for early MakerBots.

As the tools continue to become more affordable (I will go over this more in Chapter 6), the upfront financial investment will continue to decline. Especially when you're spreading the cost over a community of excited users, they will more than pay for themselves in enjoyment and learning.

Step 4: Administrative stuff

Not far down the path of starting your own makerspace, you run into the big administrative issues: insurance, liability, membership dues, etc. Luckily, you're not reinventing the wheel. Groups like Artisan's Asylum have already

---

2. Resources for makerspaces in libraries (*http://bit.ly/16GVD2l*).

done a lot of the heavy lifting in that department and many of those spaces are eager to share what they've learned. Ask a nearby makerspace, or makerspace model you admire, if they have any tips or resources before you go about the process of creating a space yourself.

More Makerspace Resources

- Makerspace Playbook (*http://bit.ly/166cwNO*)
- Gui Cavalcanti's "Making a Makerspace" entries (*http://bit.ly/1bPyrhV*)
- Eric Michaud's "How to Start a Hacker Space" series (*http://bit.ly/17H3VDT*)

Tool Lending Libraries

All these new models of maker- and hackerspaces are great, but they do fail to address the obvious question that arises when you talk about access to tools: what if you need the tools at home?

After all, the romantic idea of the maker culture is garage creation. Some projects, like home remodeling or modification, are inextricably linked to a specific place. Other projects are just too big to get out from your backyard, like Doug and Kay Jackson's recreation of a 1894 wooden submarine called the Argonaut Jr.[3] They had built the entire plywood and epoxy contraption—14 feet long, big enough for two people and multiple scuba tanks—in their front yard in Tulsa, Oklahoma. It didn't move from there until they were ready for a test run in a nearby lake. It's hard to imagine dragging something that large into a community space every day. And there are probably a million other projects and reasons that you need to use a specific tool at your house or somewhere outside of a makerspace.

One emerging option for home-based projects are tool lending libraries, which operate just like public book lending libraries. Still an infant idea and concept, there are less than fifty of these tool lending libraries scattered around the country. They differ somewhat in size and structure—some are public whereas some are privately

---

3. You have to see their wooden submarine for yourself (*http://bit.ly/155oLhj*).

owned. But they're all aiming at a similar target, and each is creating a vital resource in their respective maker communities.

Luckily, one of the oldest, the Berkeley Tool Lending Library, was right in my neighborhood. It was created in 1979 and, despite my ignorance, has been a useful and beloved asset to the community ever since. Peter McElligott, the primary staff member of the library from 1979 through 1999, recounted some of his experiences of the tool library in an interview with Jonathan Gray (*http://bit.ly/1anSiXz*). Here is how McElligott described the diverse uses the patrons have found for the tools and library:

> *People are always doing stuff in their homes. A lot of people have ongoing projects and you see them every few days for a month or two, and they're gone and you don't see them for six months or a year and then they're back doing something else... There are a certain number of street sellers and people who make craft items that they sell. There was one guy who made African instruments—kalimbas, shakers, and stuff out of gourds. He sells them at local craft fairs. He's been using the tools for years. There are various artists that use the tools for their work. There are a lot of people who I have no idea what they are doing.*

I had to check it out for myself. I casually mentioned my plans to visit the tool lending library to a friend of mine, Stacy, who also lives in Berkeley. Unfortunately, she was busy at work so she couldn't join me, but she did inquire about what I was building.

"Oh nothing in particular," I responded. "I'm just really interested in the concept and wanted to check it out. Why? Have you heard of it?"

"Of course," She said, matter-of-factly. "I use it all the time."

"Really?" I was curious. I knew Stacy was a handy person and was always working on something, but I was still surprised to hear that she used the tool library. As it turned out, she actually *was* a tool library regular. She finds herself there every few weeks, everyone knows her name, and her picture is even on the website! She's used the tools to make shelves, build bookcases, and fell trees in her backyard. Her latest project is a canoe paddle she's crafting in her backyard.

I was amazed. Even though I had read that the library was used heavily, for some reason, hearing it from Stacy made it seem more accessible.

When I finally made my trip to the library, I ended up arriving early, about ten minutes before their opening time at noon. I wasn't the only one there, as two other cars had pulled up to the library door. A man got out of his truck and started unloading a number of different tools. One by one, he set a saw, a hoe, a gardening rake, and numerous other tools next to the door of the still closed library. I asked him what he was building, and he excitedly explained how he had turned most of his backyard into a raised-bed garden. He called me over to the truck, pulled out his phone and showed me pictures of his new home agriculture operation. He couldn't hide his excitement and sense of accomplishment. It was infectious.

By the time the library opened, a few more people had shown up and it was abuzz with activity. It wasn't what I had imagined. I think I had pictured more of a traditional library with tools on the shelves instead of books. With the doors now wide open, it revealed the library to be just a desk with a long storage shed behind it. The shed was filled with tools: overflowing shelves, packed drawers, and something hanging on every square inch of the wall. The one staff member began rolling out wheelbarrows and opening other doors to sheds behind me that I hadn't noticed. The person from the other car set a large table saw down on the checkout desk. Another patron waited at the desk for someone from the staff to help him. As I looked around the library and snapped photos of the sight, the man in line noticed my curiosity and commented, "You'd be amazed how much they have tucked into the nooks and crannies back there. They've got everything."

More important, it has a staff that can help you find exactly the tool you need. The Berkeley Tool Lending Library is a vibrant community resource. For a new maker—someone who needs access but isn't ready for ownership—this is an incredible gift.

START YOUR OWN TOOL LENDING LIBRARY

My experience in Berkeley reminded me how lucky I was to be living in the Bay Area, where I have easy access to places like TechShop, Noisebridge, and the Berkeley Tool Lending Library. I had been living in a bubble. This was the easiest place in the world for someone like me, with no experience, to get started making. With a little initiative, the infrastructure was already in place for me to learn. But what about other cities that didn't have a TechShop or Noisebridge or a city-funded tool lending library?

I started doing research into other tool lending libraries around the country and how maker communities were self-organizing outside of the Bay Area. One of the first people I spoke with was Gene Homicki, one of the organizers of the West Seattle Tool Library. Gene's background was in technology. He owns and runs his own technology consulting business, which offers services ranging from custom software development to general technology strategy. Gene's other passion, as he explained to me, is sustainability, which spurred his interest in the tool lending library. Inspired by what was happening in other cities around the concept of collaborative consumption, Gene and a few others decided to take things into their own hands. After putting the idea for a tool lending library out to local sustainability blogs, they got an incredibly positive response. Before they knew it, the idea had a lot of momentum and a number of excited volunteers. Even though they had a lot of enthusiasm, there were still a number of hurdles that they needed to clear: insurance and liability issues, figuring out an inventory management system, building a bigger community around the library, and actually finding a physical space for it.

Pretty soon, the pieces began to fall into place. Gene—seeing an opportunity to combine his tech skills and passion for sharing and making—began working on a software system to manage the tool inventory. By the time the West Seattle Tool Library opened to the public, Gene had created an incredibly robust system to manage and automate not only the tool inventory, but also the membership, tool maintenance, and check-out times. By Gene's own admission, the West Seattle Tool Library boasts the most sophisticated system in the country, which supports their 1,500 tools and 400 members.

This turns out to be great news for all of us because Gene decided to offer the software to other tool lending libraries to use. The service is currently in place at the Vancouver Tool Library, with others planning to adopt it soon. You can learn more about the specifics on their website, Localtools.org (*http://www.localtools.org*).

Starting a tool library is more than just creating a collection of tools—it's organizing a maker community. It's a perfect stepping stone to creating a full-on makerspace. You can get started with just the tools sitting around in your garage, and collect more as funds allow or others donate to your cause. Combined with organizing a maker meetup, setting up a basic tool lending library is a great first step to catalyzing the maker community in your area.

Sharing as a Resource vs. Sharing as a Strategy

Of course, starting a makerspace or a tool lending library is great for the broader maker community, but how does it help you? How does it get you any closer to your own maker goals?

If you're still figuring out what your big project will be, spending time building the maker infrastructure in your community is a great way to build a strong foundation for future projects. As we'll explore more in Chapter 7, these spaces and communities are the fertile ground from which ideas can sprout.

It's also important to think about these sharing models as a strategy to help defray costs for the tools you need. Like the early NYC Resistor members who shared space and all chipped in for a laser cutter, the communal use allowed for MakerBot and dozens of other projects to get off the ground. Artisan's Asylum allows members to donate tools to the space in return for reduced membership costs. Using the different sharing methods might just be the best way to afford that tool you've always wanted or needed.

But What Tools? Finding a Maker Personal Trainer

I would have been completely lost inside TechShop (and probably wouldn't have come back out of sheer intimidation) if it weren't for a conversation with Zack Johnson, my designated Dream Coach at TechShop. As a new member, I was encouraged to set up a meeting with a Dream Coach to better understand and define my goals. With the preface that I was a complete beginner, I explained the Open-ROV project to him, along with the bombshell news that we hoped to have it all working and ready to go before World Maker Faire in New York City (only a few months away).

"So do you think it's possible?" I asked with a slightly concerned, yet curiously hopeful, look on my face.

"Honestly, not really," replied Zack. "But I do think we're going to give it a good shot and we're definitely going to learn a lot. Here's how we can think about it..."

Zack's background was in electrical engineering, and like all the Dream Coaches at TechShop he was handy with just about every tool there. If he didn't know how to do something, he knew exactly who would. Until that point, OpenROV had been mostly the brainchild of Eric, but his mechanical engineering wizardry could only get us so far. In addition, my lack of technical-anything was not contributing at all. To cross the gap to a working prototype, we needed an outside perspective and I needed some basic education. Zack understood both of those challenges and together we mapped out a plan.

During that first meeting, in addition to the background of the project goals and technical issues (as best as I could understand them at the time), I also asked him what skills and tools I needed to learn. His guidance was exactly what I needed. Zack didn't just tell me what needed to be done in terms of tool training and classes, he told me why. He reinforced the importance of the process, and assured me that it would be just as rewarding as the end product. We broke the action items into two categories: classes and workflow. Classes I should take, and workflow for attempting to have a prototype of OpenROV working for Maker Faire. The class schedule included Intro to the Laser Cutter, Basic Electronics, CNC (at the time, I had no idea what that stood for), CAD software, Arduino, Silicone Moldmaking, and a few others. As excited as I was about the classes, the workflow plan really helped to instill some much needed confidence. Even though Zack hadn't had the chance to talk to Eric, his methodical thought process of how to tackle the challenge was helpful for me to hear. It got me thinking about how the different aspects of the robot could be separated into understandable and manageable pieces.

A lawyer once told me that the main purpose of law school was to get you to think like a lawyer. I suspect a similar truism exists for engineers based on the way Zack explained the process of trying to solve the ROV issues with limited information. For a new maker, it's really intimidating to walk into a shop and survey all the equipment. It becomes a lot more digestible with a short list of specific tools to focus on first. Zack's list took the weight of uncertainty off my shoulders and pointed me in a direction to go.

Even if there isn't someone with a formal title of "Dream Coach," it doesn't mean that there aren't experienced makers who are willing to give some guidance. Often due to their grassroots organizational structure, many hackerspaces don't have a program or system to help and welcome new makers—it's almost always an issue of resource constraints, never malevolence. However, it's easy to take your education into your own hands.

Asking for guidance or project advice is a great way to get to know the people and the tools at your local maker community. Once you find a nearby makerspace or maker meetup, finding the right experienced maker isn't too challenging. Ask around for a veteran member, and hint that you're looking to get some advice about a project you're working on.

I suggest asking for a list of five classes, skills, or tools that you should focus on. I've found that five seems to be pretty close to the right number; it's at least enough to give you a broad swath of experiences and certainly enough to keep you busy for a few months of part-time work. Also, ask for introductions around the space to experts (or "know-enoughs") in the areas you're hoping to learn. After all, the knowledge and know-how resides with the people, not in the tools themselves.

It's worth restating: the goal is not to master any of these tools. The aim is to expose you to the potential possibilities by understanding the different resources. And most of all, to begin to build your own network and your own unique maker community.

Let Your Workshop Build Itself

Despite how wonderful and effective the new shared maker resources are, there is still something magical about your own personal workshop. I didn't realize it until I had a space of my own.

It was one of those moments that catches you off guard. My workshop, the OpenROV HQ, had evolved into a wonderful little makerspace of its own. It's located in a small R&D laboratory in Berkeley, California, within walking distance of where my boat is docked in the San Francisco Bay. It's full of tools now: a full electronics bench, screwdrivers, wrenches, saws, epoxies, glues, and acrylic. It has everything we need to build underwater robots (except an expensive laser cutter, which we still use at TechShop, but we'll even have one of those soon).

I was there one evening by myself, staying late at the workshop to get some writing done, when the reality of having my very own work space actually set in. In the process of writing about the different collaborative models, I realized that despite never intending to maintain my own space, I had created a unique workshop that was uniquely suited to my particular flavor of making, and I absolutely loved it.

Utilizing the different sharing models—makerspaces, borrowing from friends, lending libraries—is the right way to get started. There's little in the way of capital investment, maintenance can be shared, and most important, it's a way to meet and get to know the maker community in your area (and around the world). However, the rise of these new collaborative spaces and options isn't eliminating the elaborate and celebrated workshops. In fact, as more people discover the joy of making, makerspaces become more crowded, and tools like 3D printers become even more affordable and easy-to-use. It seems that the personal workshop will see a renaissance of its own.

A year prior, the space that OpenROV HQ occupies would have been no use to me. Quite literally, it was a waste of space. But now, having spent time learning and tinkering, my workshop seemed like an extension of my maker self; a place where I could make progress on and complete projects as well as push the Open-ROV prototypes forward at the fastest pace possible. The space adapted to me as much as I adapted to it.

It wasn't always one location, either. It started at a table at TechShop in San Francisco when I realized I would need acrylic glue for the OpenROV and I didn't have any. It became a bag of materials and tools I knew I would need (and that I could carry). This bag became the solution to inevitable future problems. My experiences, the struggles and the mistakes, were the guide on what to include. Pretty soon, this "space" spilled over into Eric's garage in Cupertino. The space became our own, filled with everything we would need or might need on our mission to make a better underwater robot. After our Kickstarter campaign, the garage began to fill with boxes of robot parts for kits we were sending all over the world. A section of the garage became the shipping department, with all of our boxes, a scale, and shipping materials. It wasn't long before Eric's roommates had had enough, and we went in search of a bigger home. We found an ideal location in Berkeley. It was the right amount of space and conveniently located right on the water in an area called Aquatic Park, where we could test our robots right outside the building. We organized the space to suit our needs, accumulating more tools and each time building a better system for keeping track of what we had. Our space, shown in Figure 4-1, is as much of a work in progress as the robot itself. They are a reflection of each other as well as an indicator of our growth as makers.

Figure 4-1. Our space

Before you go out and build an addition to your home or clear out the garage to accommodate a workshop, here are a few things to consider:

Take the low road

In his book *How Buildings Learn: What Happens After They're Built* (Penguin Books, 1995), Stewart Brand identifies a particular type of building that he calls "low road." He describes the cheap and flexible spaces that are often overlooked, such as shipping containers, garages, and old warehouses. These are the types of buildings and structures where no one cares about punching a hole in the wall. The result of this "low road" mentality creates a space of maximum adaptability. As you grow as a maker, this flexibility to change is all-important.

Makerhoods

Our OpenROV space is great. It's perfect for us. However, we still rely on the laser cutters and CNC machines at TechShop. The reason we're able to survive in our own space is not because we have everything we need, but because we're part of a growing network of nearby makers and maker businesses. It's the larger tools at TechShop, the advice and support of friends we've made through the Hardware Startup Meetup, and the screw-sorting volunteers (basically a group of our friends). It's our Makerhood.

It goes beyond neighborhoods, too. Entire cities are getting on board with this idea of making things again. Here in the Bay Area, a non-profit organization called SF:Made is organizing and lobbying to make the city a more desireable and effective place to manufacture. After successfully supporting and elevating the local community, the group is now shifting its focus on helping other cities and organizations pursue similar initiatives.

Making doesn't happen in a vacuum; it's the product of a supportive community. Another benefit of slowly growing your own workshop is that you'll spend more time integrating and using the community resources.

Take your time, maintain the tools, and stay organized

Especially if you're brand new to making, don't rush the creation of your own space. You don't need to go out and buy a new 3D printer right away. Play the long game. The workshops with the most character are those that have evolved over the years. Acquire the tools when you need them, but approach each need as a problem you want to permanently solve. Look at each tool, even the screwdrivers and wrenches, as an investment in your evolving workshop.

Build a system that helps you organize your materials. As a maker, you're only as effective as the tools you can use (and, more important, find). You'd be surprised how easy it is to avoid sorting resistors because you're not sure when you're going to need them, only to kick yourself later when you have to dig through a pile of parts to find the one you need. Any time you put into organizing will save you later on.

DIFFERENT TOOLS FOR DIFFERENT TYPES OF MAKERS

"Would you tell me, please, which way I ought to go from here?"

"That depends a good deal on where you want to get to," said the Cat.

"I don't much care where..." said Alice.

"Then it doesn't matter which way you go," said the Cat.

"...so long as I get somewhere," Alice added as an explanation.

"Oh, you're sure to do that," said the Cat, "if you only walk long enough."

— LEWIS CARROLL

Alice in Wonderland

Knowing what tools to learn and use is a big question. Bigger than just the few pages I've written and certainly more complicated than having someone suggest five tools to learn (although that's a good place to start). It always depends on what you want to make. In broad terms, it depends on what type of maker you want to become, which can bring you back to the same intimidating starting line.

I get that. I understand the need for direction, but I don't think there's a map. In fact, the closest thing I've found is more of a compass, something to hold onto when the surroundings are uncertain. It's the idea that somewhere along this path —if only I walk far enough—I'll find my own craft.

5

Craftsmanship

Craftsmanship was a simple idea. Before I started my maker journey, I thought it was something I could easily identify and define. In fact, it was the lack of craftsmanship in my own life that was the real, underlying motivation for getting involved. I'll never forget the feeling of maker admiration when I attended my first Maker Faire, or the conversation I had with the carpenter in the days after losing my job. These people (both men and women) were craftsman. They had an approach to their work that nobody—not an employer or an angry customer or a turbulent economy—could take away from them. It was deeply personal and, from my perspective, fulfilling in a way that I had never felt about my work.

Strangely enough, the idea that seemed so certain when I began my journey became the most challenging and elusive concept I tried to tame. Initially, the concept of craftsmanship seemed like a title. It was a badge of honor that could be achieved, like being a doctor or scientist. I tied the word to the traditional tools, the ones my grandfather and grandmother used in woodworking, metalworking, or sewing. I tried to spend time with as many craftsman that I could, from an old-school metalworker in the foothills of the Sierra Mountains to a bag maker in the Mission District of San Francisco. I wanted to get a flavor for the analog tools—the skills I missed in shop class—but also soak up as much wisdom as I could from this dying breed (or so I thought) of worker.

It seemed straightforward: meet and learn from experts in a few different trades, take notes, and hopefully walk away with a little more tool knowledge and wisdom. But the further I went down that path, the fuzzier the concept became. I realized that craftsmanship wasn't a title, it was a way of moving through the world. It was as much about learning the techniques as it was about discovering yourself. I also realized that craftsmanship wasn't confined to the classic trades I had originally envisioned, but spanned all types of tools. And the same forces that were shaping the maker movement—the accessibility of new tools, the connectivity of the Web, and the DIT mentality—were forging a new model of a craftsman for the twenty-first century.

The Process of Becoming

In my search, no one embodied this new class of craftsman more than Joel Bukie-wicz, a knife maker in Brooklyn. I learned about Joel, as I did many others, after watching a short 10-minute documentary about him, his craft, and his perspective on making. The video was shot as part of the Made By Hand series (*http://vimeo.com/31455885*). It consisted of Joel talking over clips of himself in his gritty workshop in the Gowanus neighborhood of Brooklyn. I was captivated by the visual tone of the film and of Joel's world.

Like for me, making wasn't always a part of Joel's life. It was something he stumbled into during a particularly tough and uncertain time. Joel was living in Brooklyn, having just graduated with a degree in fiction writing from the New School, and was trying to get his writing career off the ground. After a prolonged struggle to sell his first manuscript, fear and doubt began to set in. It became so worrisome that Joel decided to take time off from writing. He let go of the fear by letting go of the hope.

With the hole that writing left, Joel began to fill his time with new "creative offerings" like making bookshelves, canoe paddles, or fixing things around the house. While spending time in Georgia, he began teaching himself how to make knives in a garden shed behind the house. The process resonated with him. The making of knives had the feel and fundamental utility that pushed all the right buttons. He started with sporting and hunting knives, putting thousands of hours into his newfound craft. After coming to the realization that the knives weren't actually being used, and valuing the idea of utility more than art, Joel switched his focus to creating knives for the kitchen. His passion has since turned into a thriving small business in Brooklyn, where Joel still makes every knife by hand.

I've never made a knife. But I'd like to try. I wonder if I'd get that same rush, the internal "click" of something that feels right. That was the dream when I started. I wanted to find something that I could pour myself into. My starting point was the same as Joel's: a creative hole that I needed to fill with something tactile. Joel's search, moving from bookshelves to paddles, mirrored my own quest, which was a sampling of maker aesthetics. Like finding the right pair of jeans, it involved trying on a number of different roles, sitting in them for a moment, and trying to get a feel for whether this was "me." Only after you have something that fits can you start breaking it in.

It wasn't until much later, after I was knee-deep in underwater robotics, that I realized how critical that search had been. I learned that craftsmanship is not a destination, it's a process. And the search—the trying-on process—was the first phase. It's impossible to achieve mastery without starting the search. It feels similar to unlocking a padlock: twisting to the right, then back to the left, then back to the right. That doesn't work. Back to twisting. Then, just when you're ready to break the damn thing, something clicks. All the frustration and failed attempts are behind you, and were, in fact, part of the unlocking.

A Maker Sampling Quest

You picked up this book for a reason. Something about the idea of creating an actual, physical object spoke to you. Based on my conversations with other reluctant makers, this could be that you had a product you have always wanted to create. Or maybe it's that you read about 3D printing and wanted to see what all the fuss was about. Fundamentally, though, I'm willing to bet that there is a part of you that sees this maker movement as scratching an itch. It seems to me that that itch lives somewhere near this idea of craftsmanship—tactile and meaningful work—that I just described.

If you don't have a specific project in mind already, but you find yourself driven by the same search for authentic creativity, I encourage you to recreate a similar "Maker Sampling Quest" that Joel and I found ourselves on (Joel's writing sabbatical and ensuing "creative offerings" and my Zero to Maker journey). There's something re-humanizing about the process. For me, it became very clear that human beings are tool builders and tool users; it's ingrained in each of us. It creates a connection to the past and a sense of responsibility to the future. Committing myself to the month-long search helped me to crystallize that perspective.

The natural inclination is to rush the craftsmanship process, to squeeze into the first maker aesthetic you find. I think it's worth it, if you can afford it, to designate the first month as a re-skilling exploration.

Here are some ideas and tips:

Include the old-school tools

It's easy to want to jump right into the new tools like the laser cutter and 3D printer, and you should (we'll explore that more in the next chapter). The ease of use and the immediate feedback loops of the digital fabrication tools are great for new makers. But I also think it's important to become familiar with the manual processes, too. Stay curious about classic tools, even if you have no idea how or when you'd use them.

For one, it will give you a greater respect for the newer tools. Learning to make a mold out of wood by hand will help you appreciate the speed of the digital substitute of laser-cut cardboard and Bondo (a two-part putty use for body filling). It will also give you a better sense of what you're gaining or losing in quality.

More important, though, you're probably going to need to use them. As incredible as a CNC mill machine can be, it's very likely you'll have to do some touch ups and sanding using analog tools afterward. And sometimes a table or miter saw is the fastest and most effective tool for the job. The digital tools aren't replacing the traditional tools, they're enhancing them.

Meet the old-school people

If my grandfather were still alive, and I told him I was getting into "making" he would look at me with a curious and concerned look. Of course you're going to make things. What else would you do?

There is an entire generation that feels uncomfortable *not* making. The wisdom that these makers have accumulated in their hands, through a lifetime of dedicated practice, is something that will never be available through an online course. It's a type of knowledge that can only be absorbed by spending time around them, watching the subtleties and doing your best to imitate. Even a day or two with this class of maker can be revelatory.

Make it a phase

Originally, my pitch to the *Make:* editors was "Zero to Maker in 30 Days." I did that for a few reasons. The first was to set a deadline for myself, which was an attempt to stave off procrastination. But it was also because I didn't have that much time. Having just lost my job, I had a short window of time before I burned through all of my savings. I thought it would be a short jaunt and then back to finding a job.

I think this designated exploration period was critical. It kept me from getting overly invested in the first project that I encountered, and it forced me to gain a broader perspective on the possibilities of the maker movement. It also caused me to go outside of my comfort zone and meet makers whom I might not have met otherwise. It's been surprising how often I've resorted to some piece of advice or tool I never thought I'd use.

Setting a distinct period for exploration, say two or three months, is a great way to start.

Note

Now that you've made it this far into the book, I should probably clarify a few things. Mainly, I feel I need to issue you a warning; you deserve an explanation of what this book *isn't*.

As much as I would love this to be a comprehensive guide to every tool and technique in the maker universe, it's just not possible. For one, the idea of making is just too broad (and too important) to fit within one book. It's everything from knitting and wood carving, to waterjet cutting and brewing beer. Makers are everywhere—every person who is out there changing and affecting their world in a physical way is a maker. I could never hope to explain all of it. And I wouldn't want to. In fact, it's the endlessness that makes it so rich.

Nowhere were the limits of my own perspective more obvious than with the craftsmen I met. There's only so much one person can see or do in a year, and unfortunately, I didn't have nearly enough time to dive into the deepest depths of craftsmanship. That takes a lifetime. This book is a small window into a big world.

Admittedly, this chapter is sorely lacking in specifics like how to use a bandsaw or how to make your metal shop from scratch (for that, Google "Dave Gingery"). It was an issue I wrestled with for a long time. Instead of putting down lists and links, I decided to focus on the ideas I discovered around craftsmanship.

If you are looking for more specific details on a tool or technique, refer back to the Ultimate Maker Tool in Chapter 4.

Sometimes It's Simple: Make What You Know

As much as I talk about this being a journey into a new world, don't dismiss the knowledge and materials in your immediate surroundings and past experiences. In fact, after you start looking, it's easy to find the maker spirit everywhere around you. Sometimes you don't have to go far to find the feeling of craftsmanship. Sometimes it's been with you all along. It just might be a new way of approaching what you already know. Susan Hoff taught me that.

Susan makes bags—beautiful, durable bags handcrafted from old sails and horseback riding reins. By grabbing a seat in her small studio in the Mission District in San Francisco, I had an inside look into the entire sourcing, manufacturing, marketing, and accounting departments of Susan Hoff, Inc. A large, sturdy table with a Sailright sewing machine takes up most of the room. Finished bags, about eight in total, adorn the north wall while shelves stuffed with old, folded sails and reins fill the rest of the space.

"I think this is the smallest industrial sewing maching you can get. It's made for sewing sails on board boats," she explains. Clearly, she doesn't need much. The Sailright and an old-school push palm are the only tools she uses.

The bags, the tools, the studio: it all seems to fit together. It's just so unapologetically her own. And the more I learned about her background, the more it came together.

Hoff had a hands-on childhood. When she was in middle school, her family bought an old, condemned farmhouse in northwestern Illinois. They spent the next five years completely renovating and rebuilding. She credits the experience as giving her the confidence to use tools. It was also during those high school years that her mom first taught her how to sew.

In addition to life on the farm, she spent time sailing at summer camps, eventually becoming an instructor at one of them. Her love for boats and the water grew when she spent a semester of college sailing around the Caribbean on the 134-foot schooner Corwith Cramer. After college, she spent more than three years leading sailing trips, some of them 3 or 4 weeks long, for Outward Bound.

The bag business was almost an accident. While working for Outward Bound, someone gave her an old sail that was well beyond its useful sailing days. She loved the texture of the canvas and decided to try and make a bag out of it. The results were great so she made another. Then another. Pretty soon she was getting requests from friends for bags of their own. She told me she had a breakthrough when she realized she could make money selling her bags in a few small retail stores in Maine. The bag business has grown from there. Whenever Susan has traveled or moved to a new city, she's brought along her Sailright and set up shop.

She's currently thinking about opening up a retail location in San Francisco, but even as we talk about it, she seems to go back on the idea. It's not that she doesn't think it would succeed—she's confident it would. She's just hesitant about the commitment. She'd rather take two months off and sail across the Atlantic Ocean (which she's done) if the opportunity arises.

It's hard not to be charmed by the life Susan has created for herself. The simplicity is alluring: taking old, undervalued materials and working them into a new, useful life. The value is so tangible. She gets out exactly what she puts in.

I've grown to love that quality of the makers I've met. They create value. It's not an indifference to the grander scheme of things, but a sufficient preoccupation with immediate possibility. They don't seem overly concerned with changing the world. They're content to find a small corner of it—a place that not enough people are paying attention to—and make it beautiful.

It doesn't necessarily take a 3D printer or an Arduino microcontroller to start making things. Sometimes the project you've been searching for—your maker sweet spot—has been a part of you all along. Susan is a perfect example. She didn't have a blueprint for her maker life and business. She just started making what she knew.

Down the Rabbit Hole: Welding

When you venture into the unknown (and you should) with skills that are unfamiliar, you learn that all of these tools have their own unique histories. Even the latest and greatest have a maker lineage that can be traced back through generations of tool users. Sometimes there is new terminology or slightly different styles and techniques, but there are always people who came before. We're all standing on the shoulders of giants.

Part of learning a craft is learning that history. This became evident to me as I tried my hand at welding.

I actually had no good reason to weld anything. There were no parts of the ROV that needed to be welded together. I didn't have dreams of building my own grill or fixing up an old motorcycle. There was just something about joining two pieces of metal together. For me, it was a romantic idea about welding masks, torches, and flying sparks. I had to try it.

Like any tool in the shop, as soon as you scratch the surface, you quickly realize just how much there is to learn, and how specialized the knowledge becomes, depending on what you want to do.

MIG WELDING

My first experience with welding was taking a metal inert gas (MIG) welding class at TechShop. At this point, I knew there were different types of welds, but I couldn't tell which was which, or what they were used for. MIG welding uses a continuous wire feed (which serves as a filler to adhere the two pieces) as an electrode and an inert gas mixture (argon and carbon dioxide) to protect the weld from contamination. From what I was able to take away, MIG welding is fast and, because of the automatic wire feed, somewhat easier to learn.

Our instructor spent a lot of time on safety and preparation, which are both important aspects of welding. There were a few other students in the course, each with a slightly different grasp of what they were getting themselves into. After he set up the table, each of us were given a chance to handle the arc. Admittedly, my first welds were not very good. I was zigging and zagging all over the sheet metal nowhere near the joint I was trying to weld. It took me a while before I was used to the darkness of the mask and the feel of the torch through the gloves.

The main lesson I took away from my first welding class wasn't a safety tip or technique. It was a quip by our TechShop instructor: it takes about a mile of welding before you're decent. I did the math in my head, adding up the different activities and examples we just did in class. I estimated it was about two feet. I had a *long* way to go.

TIG WELDING

Because MIG welding had an automatic wire feed, it was easy to focus on my welds (even though they weren't very good) and the speed and angle that produced the different results. Tungsten inert gas (TIG) welding was a slightly different process. You have to feed the filler metal into the weld manually while simultaneously controlling the arc (the part of the metal being heated) created by the tungsten electrode. Even though this was a slightly more complex process, my welds were dramatically better—an improvement I credit to the MIG welding experience, and probably only because I had gotten used to seeing through the welding mask at this point. Based on my conversation with the instructor, the TIG weld can be more precise but can take a lot longer and cost more than MIG welding.

GAS WELDING

The general classes at TechShop were informative, but they left a lot to be desired. A Zero to Maker onlooker suggested I check out TM Technologies and its metalworking courses.[1] I followed the suggestion and discovered that TM ran weekend workshops on metalworking fundamentals and a four-day metalworking intensive out of Kent "The Tin Man" White's workshop in Nevada City (about three hours northeast of San Francisco). I decided to reach out to the Tin Man to see if he could manage a tour of his workshop. I sent him an email explaining what I was doing, and that I'd love to learn more about his work. A few days later I got this response:

"Come on up. I'll feed the bears first if you call ahead."

Understandably, I was a little nervous as I drove to the shop on a Sunday morning. But as it turned out, making that trip and spending that morning with Kent was one of the most enlightening experiences of my journey so far. Not only did Kent give me a lesson in gas welding (oxyacetylene welding), but he offered his point of view on the Zero to Maker concept, which was a situation he knew all too well. This was his advice for me as a beginning metalworker:

---

1. For more information on TM Technologies, check out their website (*http://www.tinmantech.com/html/ kent_white.php*).

My advice for someone getting started is to read some and watch some. Ask questions. Then decide what you want to do. Start simply. Learn to sketch, measure, mark, cut, file, and sand. Learn also to drill, deburr, fold, and bend. Learn to rivet, bolt, and screw. Learn the metals and their applications. Then learn the hot stuff, after your shop skills are developed. Nothing worse than jumping in prematurely and setting your hair alight.

The tone in his voice was unmistakable. He was someone who genuinely cared for the craft, but also full of concern after watching a generational drop-off in metal workers. He's become a world renowned expert in gas welding instruction not only because he's exceptionally great, but also by default—he's one of the few who still use and teach the technique. I asked him why he prefers gas welding to some of the more common methods I'd seen, to which he responded:

Gas welding is simple and portable and needs no electricity. Perfect cleanliness and breeze-free conditions are not required, as they are with MIG and TIG. Persons nearby do not necessarily need to be shielded from it, as they must be from arc rays. It is effective on several types of thin sheet and tubing, such as steel, aluminum, stainless, copper, etc.—and the same equipment is also appropriate for soldering, brazing, annealing, hot working, coloring, and in some cases, cutting—which the marvelous electric machines simply cannot accomplish.

The way the Tin Man talked about gas welding was very different than the way my TechShop instructors had described it. He pulled me around his workshop to show me books and diagrams, special welding glasses he had invented, and work some of his students had created. Metalworking was clearly a part of him. And he was a part of it. He was one link in a long chain of craftsman.

Twenty-First Century Apprenticeship

I enjoyed gas welding with the Tin Man, but it wasn't my "thing" like Joel's knife epiphany or Susan's bag making. It was just a memorable pit stop on my maker journey. But if I ever want to dive deeper into welding or metalworking, I'll know where to go. I would try to create some type of informal apprenticeship under the Tin Man. I would get my mile of welding in while trying to understand what it meant to be that next link in the chain of metalworkers.

Apprenticeships are an old idea, dating back to the Middle Ages, and they're still around today. All over the world, some form of the apprenticeship model has been used for centuries to pass specific crafts to the next generation. Apprenticeships are still a fundamental feature of the education system in many countries, especially Germany. They are still alive in the United States, too, but they've taken on a rather narrow role. Currently, they are used mainly by the trades and labor unions—electricians, painters, plumbers, and others—as a bridge from school to work. It's still an excellent option if you aspire to a career in the trades.[2]

However, I think there's an opportunity to apply the apprenticeship model in new, more tactile ways than the "momentary apprenticeship" of YouTube and the traditional apprenticeship of the trade associations. This model would be less formal, more broad, and more realistic for the work and life demands of new makers and experienced craftspeople alike. For example, if I decided that I wanted to learn metalworking, it's terribly intimidating for me to think about going back to school with the intention of locking myself into one type of work for the next 30 years. But I also wouldn't trust myself after watching a few YouTube videos.

Instead, I would set about creating my own apprenticeship—not based on a set curriculum or National Institute of Metalworking Standards (NIMS), but based on absorbing the knowledge of the most skilled craftsman I could find. To continue with the metalworking example, I would convince the Tin Man to let me work for him, making sure I was clearly articulating two things:

1. A deep and committed desire to learn the craft.

2. An admission of inexperience and an enthusiastic offering of some tangential form of value, like helping to blog or run social media campaigns to try to increase attendance at the workshops. I would be trading the digital currency I have spent my life accumulating for something more tangible. Or, I would do anything where I could immediately add value, even if that were just answering the phones and brewing coffee.

---

[2]. In general, trade schools are an excellent place to augment your maker education. They have programs and certificates that can put you beyond the "enough to be dangerous" level and head-first into a career. They're also a resource for taking classes and learning tools on a one-off basis. Not every hackerspace is going to have a welding setup or advanced CNC machines. Community colleges and trade schools can fill in the gaps.

It's one thing to learn how to make knives in a class at a makerspace, but it would be quite another to shadow Joel and learn how, in addition to making knives, he manages his small artisan business in the digital age; that's an art form in and of itself.

In his book *Mastery* (Profile Books Ltd, 2012), Robert Green emphasizes the essential step of apprenticeship in achieving mastery: "...the goal of an apprenticeship is not money, a good position, a title, or a diploma, but rather the *transformation* of your mind and character—the first transformation on the way to mastery."

That fits with my idea. The notch on the belt isn't as important as actually absorbing knowledge. Greene goes on to outline three phases of this self-learning style of apprenticeship:

Deep observation
Learning the unspoken rules and social dynamics of the new world or skill, watching intently for the factors and details of success.

Skills acquisition
Learning the tools, actions, and movements of the chosen skill. Greene advises keeping it simple at first, avoiding multitasking to focus on building a foundation one skill at a time.

Experimentation
Continually pushing yourself past the point of comfort. Testing your new skills and filling in the knowledge gaps.

Greene also acknowledges the point when it's time to move on, when the learning has reached a point of diminishing returns. The beautiful part about this self-styled apprenticeship is that you can then take your newfound skills to your next endeavor or adventure. You are not locked in to a foregone career path, and you are free to create value elsewhere.

The Future Is Hidden in the Past

Some part of me was always jealous of the craftsman (or what I imagined them to be). More than their manual skill, I felt that they were better prepared for the world than I was. They knew what was valuable. They built their lives on stable ground, whereas I had taken the "promising career in a growing industry" bait.

Somewhere along my path, I had confused the decline in the popularity of a traditional tool or technique with a decline in value. Still steeped in my tool insecurity, I began to hypothesize that older techniques were not only still valuable, but sometimes actually *more* valuable.

My thinking was based on an omission in one of the basic concepts taught in business schools and marketing classes around the country: the *Innovation Curve*, as shown in Figure 5-1. The Innovation Curve, or the *Diffusion of Innovations Theory*, was developed by Everett Rogers while he was an assistant professor of rural sociology at Ohio State University. He published his ideas in his book, *Diffusions of Innovations* (Free Press, 1962). He theorized that adopters of new innovations (technology, ideas, products, etc.) fell into distinct categories: innovators, early adopters, early majority, late majority, and laggards.

He based his theory on the bell curve, pictured in Figure 5-1.

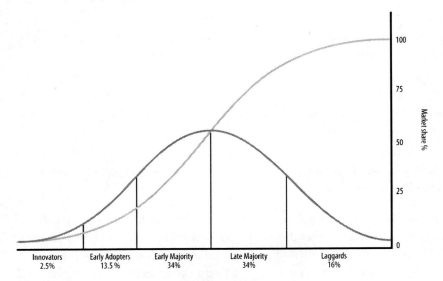

Figure 5-1. The Innovation Curve (from Wikimedia Commons)

At one point, the theory became the second most cited paper in the social sciences. The idea received a new life of sorts when it was repackaged and sold in the form of a marketing book, *Crossing the Chasm* (HarperBusiness, 2006), by Geoffrey A. Moore. Moore used Rogers' concept to explain why some technology companies fail or succeed based on their ability to attract and persuade "early adopters."

For better or worse, the terminology has stuck with us. When I first saw the Innovation Curve, I thought it was overly simplistic. I still think that's true, but have also decided that it's misleading. Not that I totally disagree with Rogers and Moore's analysis of technological innovation, I just think it only tells half the story;

it focuses solely on the rate of adoption, ignoring both the nature of use and the rate of "un-adoption."

In the real world, adoption can hardly ever be explained by a simple bell curve. Tools and technologies are continually replaced. The reasons are endless: something shiny and newer, cost barriers, unforeseen market shocks, supply issues, etc. To account for this, I've decided to create my own innovation curve, one that takes into account both the rate of adoption as well as un-adoption. The new graph, which I'm calling the *Unnovation Curve* (seen in Figure 5-2), attempts to tell a more complete story by including the general form of both technological diffusion and dissolution.

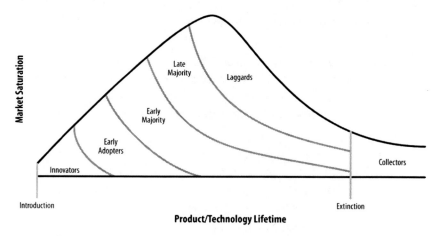

Figure 5-2. The Unnovation Curve

Is it perfect? No way! But the important part of the updated graph is to introduce and emphasize a new category of technology user: the collector. In the Unnovation Curve, a collector can come from any one of the previous groups; late adopters, laggards, early majority, late majority, and (although unlikely) even early adopters. What's important about the collectors is not when they adopted a tool, but when they refused to reject it. They are the ones left standing in the musical chairs of technological evolution.

My definition of a collector is someone who refuses to reject a technology after it has passed its peak; or in other words, it is no longer recognized as a mainstream technology. For instance, the horse as a means for transportation is no longer considered a mainstream technology for getting around. However, horseback riding hasn't gone away, it's just evolved into equestrianism. It's not something everyone does anymore, but it's something *some* people still do.

This post by David Frey (*http://bit.ly/19GzjIn*), in which he interviews farrier Danny Ward for the Tractor Supply blog, is the perfect anecdote:

> For about 30 years, Ward has run Danny Ward's Horseshoeing School in Martinsville, [Virginia], continuing a legacy his father Smokey started in 1965.

> "He shod horses before World War II, when they were used more for work and transportation," Ward says. "Then the '40s came along, and the '50s. Then the tractor came along. All of a sudden, horses weren't used for work and transportation. He kind of got out of it for a few years."

> It was a difficult time for farriers everywhere. "During the industrial era in the country, we lost the basic core of our experts, if you will, who had come over from Europe," Ferguson says. "Basically, the only horses that were getting shod during that time were those of the rich and famous."

> When the 1960s rolled around, Ward says, the outlook brightened considerably. Work horses became pleasure horses, and the equestrian industry was starting to explode. "By about 1965, he just couldn't keep up with all the work," Ward says.

> His father started the farrier school to train others to help him juggle the workload at a time when many farriers had left the industry. But with the new role of horses in society came new needs for horseshoeing. When horses were beasts of burden, farmers didn't expect much more than a regular trim of their horses' hooves and a curved piece of metal nailed in to protect the feet.

> These days, when most horses are for recreation in sports that have become high-dollar activities, the needs for horseshoes have become much more complicated. Horses may need corrective shoeing or trimming to fix chronic foot problems. A mere millimeter's width in a horseshoe can mean the difference between a horse being sound and being "off." These days, horseshoes are made to fit each individual horse and each individual hoof.

> "It's no longer a trade. It's an occupation," Ward says.

The nature and makeup of the collectors is always different, but for every technology or tool that you can imagine, more often than not you'll find a group of die-hard enthusiasts who keep it alive for one reason or another. For me, this is reassuring. I like the idea that there are people out there who are working with tools I've never even imagined. It keeps it interesting.

Take flint knapping, for instance. Flint knapping is the process of shaping stone tools for use in weapons, building, or decorations. It involves using both a hard and soft hammer to slowly remove flakes of stone—flint, chert, obsidian, or something similar—until the desired shape is reached. It's a primitive form of stone carving, and different forms of flint knapping have been used by cultures all over the world. However, after the rise of more modern metalworking technology, the practical uses of flint knapping eroded. Now the technique lives on in the hands of experimental archaeologists and a certain class of outdoorsman scattered around the world. But these modern knappers are insistent. At one point, a handful of researchers and writers even "helped to ignite a small craze in knapping" (*http://en.wikipedia.org/wiki/Knapping*) thanks to their publications.

I learned about flint knapping through the *Make:* blog and, after following a few links, quickly found myself deep inside a foreign microculture. Much of it is centered around flintknappers.com, a site that gives knappers a platform to share knowledge. The man who runs the site, Mike Miller, started it in 2001 as a way for him and a few friends to publicize and sell their work. It has since turned into the central online hub for lithic art. Mike is a lifelong knapper; the hobby stems from his interest in arrowheads as a kid. Now he runs the site in addition to his work as an archaeologist.

From my perspective, Mike Miller is an enviable entrepreneur. It's not a full-time job, but it amplifies his other work (and he explained that others do make a living as full-time knappers). He has a loyal and enthusiastic customer base. The flint knapping market is relatively insulated against market shocks, and there is no chance that a technology will come along and "replace" flint knapping. For all intents and purposes, it's holding its course. And, Mike and his fellow flint knappers are holding onto knowledge that will never expire.

In his book *Antifragile* (Random House, 2012), Nassim Taleb argues that, in fact, the technologies that have survived the longest are most likely to endure. He refers to one of his insights as the Lindy Effect: "the longer a technology lives, the longer it can be expected to live."

Taleb elucidates his theory by explaining one of his typical evenings:

Tonight I will be meeting friends in a restaurant (tavernas have existed for at least twenty-five centuries). I will be walking there wearing shoes hardly different from those worn 5,300 years ago by the mummified man discovered in a glacier in the Austrian Alps. At the restaurant, I will be using silverware, a Mesopotamian technology, which qualifies as a "killer application" given what it allows me to do to the leg of the lamb, such as tear it apart while sparing my fingers from burns. I will be drinking wine, a liquid that has been in use for at least six millennia. The wine will be poured into glasses, an innovation claimed by my Lebanese compatriots to come from their Phoenician ancestors, and if you disagree about the source, we can say that glass objects have been sold by them as trinkets for at least 2,900 years. After the main course, I will have a somewhat younger technology, artisanal cheese, paying a higher price for those that have not changed in their preparation for several centuries.

This simplified narrative is true for most of us. We're surrounded by countless technologies of the past, yet we've become blind to their provenance and oblivious to the potential of actually making them ourselves. But just because we don't notice the changes as much as the latest upgrade on our smartphone doesn't mean there isn't an opportunity to explore.

Craftsmanship Re-Imagined

Even though the past is still with us (and still valuable), that doesn't mean that only traditional methods hold the mark of craftsmanship. I came to realize the concept was much broader. It didn't necessarily equate to age or have to involve a specific method or technique. More than that, it was an ethic, a way of approaching the work.

In the moments of technological transition, when the boundaries between manual and mechanized labor begin to blur or change, an uneasy tension develops. Whenever machines start doing our "jobs," a group of people revert to a nostalgia for days and ways gone by, while others start tinkering with new opportunities, searching for a new maker aesthetic. The futurist Paul Saffo calls this the divide between the Druids and the Engineers (*http://edge.org/response-detail/23858*). It's something I've struggled with myself as I met the torch bearers of traditional skills while at the same time reveling in the immediate feedback of the newer tools.

And the divide is nothing new. In *American Genesis* (University Of Chicago Press, 2004), Thomas Hughes' account of technological innovation's effect on

American culture around the turn of the twentieth century, he describes the dynamic perspectives and outcomes from the previous industrial revolution:

> With the rise of factory production and the displacement of human skills by machine, numerous social critics lamented the passing of the era of the craftsman. In England, William Morris celebrated the joy of work and called for the recovery of medieval crafts... But in the industrial era tool users were giving way to machine tenders. Still, in the model rooms, laboratories, and machine shops used by independent inventors, craftsman of a conspicuous skill thrived. Kruesi, with an intuitive grasp of Edison's three-dimensional concepts, presided over the machine shop at Menlo Park. He transformed a quick sketch of Edison's into the first phonograph... Sperry attributed his company's success in manufacturing the precision gyroscopic devices to the skill of his machinists, many of them Swiss.

My nostalgic sympathizing with the Tin Man (and many others) about the decline in skilled machinists and my own lack of manual literacy seemed eerily similar to the tone of William Morris. But looking back, even the skilled machinists once seemed a distasteful transition from what was considered good and fulfilling work. Similarly, the migration to the digital fabrication tools doesn't spell the end to craftsmanship. Rather an adaptation of it, driven by swaths of opportunity that the capital-intensive factory production model can't serve efficiently.

If anything, the current trends of workmanship are pointing toward a more human-scale marketplace, a throwback to the small, local, and personal artisans and makers that existed before the factory production model took over. As people gain access to the powerful new tools of desktop prototyping and production, they are making everything they can imagine. And the new tools of distribution, fueled by the Internet and online communities like Etsy and Kickstarter, have made it easier than ever to connect and find a community of people to sell and share their wares.

Chris Anderson, the wisest oracle of long-tail maker economics, articulated it best in *Makers*:

These niche products tend to be driven by people's *wants and needs rather than companies' wants and needs. Of course, people have to create companies to make these goods at scale, but they work hard to retain their roots. Such entrepreneurs often state that their first obligation is to serve their community, and to make money second. Goods made by passionate consumers-turned-entrepreneurs tend to radiate a quality that displays craftsmanship rather than mass-manufactured efficiency.*

Meet the twenty-first century artisans. They understand the value they are creating. It's tactile. It's real. They made it because they wanted it themselves. They can tell you exactly how everything is made and where their materials come from. They blend the proven tools of the past with the current tools of today, picking and choosing whatever suits their aesthetic. Joel uses Twitter and Vimeo to promote himself and his knives to his community. Mike Miller has created a website as a platform for distributing flint knapping knowledge and products. Celia uses CNC mills to aide her craft and amplify her production capabilities. They're filling unique niches for the makers and users of things. It's an adaptive radiation of cultural entrepreneurship.

The new maker movement is an opportunity to discover, define, and share your own unique craft. The spoils will go to those who can find that elusive sweet spot between personal satisfaction, a dedicated approach to the work, and an openness to the evolving technologies.

Now, let's meet some of these new tools.

6

Digital Fabrication

Bilal Ghalib doesn't do handshakes, just hugs. The first time I met Bilal, at the first Maker Startup Weekend (a weekend-long making event at TechShop in San Francisco), he came up and gave me a big squeeze.

We had traded a few emails and a brief phone call in the days prior, but had never met in person. I was organizing the event and was introduced to Bilal as someone who might be a good candidate to teach the 3D printing class we were offering to participants. He seemed like a nice, gregarious guy, but I was still a little shocked when he showed up with a bright purple blazer and rainbow-dyed hair. As it turned out, Bilal's outgoing and inimitable style made him the perfect person to teach 3D printing—instantly disarming the nerves of beginners like me.

It was perfect timing for Bilal. He was on the tail end of his "Pocket Factory" tour, a road trip around the country with his best friend, Alex Hornstein, and a MakerBot Thing-o-Matic 3D printer in the back of their Toyota Prius.[1] The trip was an experiment. After spending time at MIT, Bilal and Alex had seen and imagined the potential of desktop manufacturing, and were curious to find out what the technology meant outside the media lab and in the hands of everyday folks. They drove across the country, stopping everywhere they could: schools, museums, hackerspaces, and even supermarket parking lots. They talked to anyone who was interested. And they experimented, attempting to understand what it meant to have a micro-factory in the trunk of their car.

At first, they tried selling the parts that they printed, fancying themselves a tiny, portable store. That didn't work very well. They found that people didn't value the small plastic trinkets they were creating. Then, they tried selling customized products for the "customers" they met, letting the potential customers modify and alter designs before they were printed. That wasn't profitable, either. By the end of the trip, they realized that the best way to earn money with a desktop 3D printer (and

---

1. You can find the full Pocket Factory story on their blog (*http://pocketfactory.org/*).

thus keep gas in the tank for their adventure) was to teach classes on how to use the device.

The final stop of the Pocket Factory tour was San Francisco and, coincidentally, our Maker Startup Weekend. I was a few months into the Zero to Maker journey at that point, but I still hadn't had the chance to play with a 3D printer. The MakerBot at TechShop had been out of commission since I began taking classes there. Now, with Bilal here to teach a class, and his 3D printer under his arm, I would finally get the chance to try it.

After months of being on the road and explaining 3D printing to people who had never heard of the concept, Bilal knew exactly where to start. He began with the analogy to regular inkjet printing. Explaining that 3D printing was very similar, except that instead of putting ink to paper, the machine extruded a thin layer of plastic, and then another layer on top of that, slowly building up a three-dimensional figure. He understood the challenge of beginners like me to think (and design) spatially. Instead of working with 3D models, he encouraged us to start with a 2D outline, like our name, and extrude it (pulling the design into three-dimensional space). His explanation and teaching style were perfect for the crowd of newbies. He wasn't just explaining how to work the machine, he was teaching us how to wrap our minds around the new design possibilities. He continued his lecture on extrapolating 2D familiarity to include shapes, and eventually merging shapes to create more complex 3D models, as shown in Figure 6-1.

I was excited to try my first print. I used Autodesk 123D (more on this later in the chapter) to mock up a simple, box-like container for the OpenROV, hoping to create an easy way to adjust ballast of the device on the fly. Then I sat down next to the MakerBot and pulled up the ReplicatorG software (the software that runs the printer). I adjusted the size of my design, scaling it down after re-evaluating my original shape, and then clicked print. The machine began to buzz, and my anticipation started to build. My confidence and excitement that I was able to create a physical, tangible object from digital bits grew with every layer of plastic that was laid down. The first attempt wasn't very good—it was disappointing, actually. I ended up stopping it mid-print after the part lifted off the build platform. The second time was the charm.

Figure 6-1. One of Bilal's go-to prints is the "Bunny Chair"

I carried the tiny plastic object around, showing anyone who would pay attention. Nobody cared. For anyone who had 3D printed before, my creation was nothing impressive. For anyone who hadn't had the experience, it seemed equally uninteresting. The actual object wasn't the point, though. This experience was my first taste of 3D printing and the uniquely intriguing power of bringing digital imaginations into the real world. It's impossible to play with one of these devices and not have your mind wander into thinking that, pretty soon, everyone will want to experience this act of creation.

In my opinion, it's the personally empowering act of creativity that is driving the maker movement. It's just plain fun to do this stuff (see Figure 6-2 for an example)! But with a slightly broader perspective, it's also possible to extrapolate the impact of ubiquitous digital fabrication and the impact that could have on manufacturing and the economy.

The best theory on where this is going and what it means is packaged in Chris Anderson's *Makers*. Digital fabrication coupled with the democratized distribution channels on the Internet (i.e., Kickstarter or Etsy) level the manufacturing playing field. In the same way that the Web made it possible for anyone to create and share media (movies, music, and books), the barrier for micro-manufacturers is essentially eliminated. Suddenly, we all have pocket factories.

It's your turn to create. But first, the rules and tools...

Figure 6-2. Bilal's Bunny Chair becomes a real thing. You can find the design on Thingiverse (http://www.thingiverse.com/thing:18512).

It Starts with Digital Design

Digital fabrication starts with a digital design. This was a huge worry for me. I had heard about and seen CAD programs before, and I had watched as designers and architects used them to create elaborate and complicated designs like motorcycles or buildings. I worried that I would need another degree before I could find any use for that skill. But a few weeks into my journey, I realized it was an unavoidable obstacle if I wanted to play with the digital tools. Luckily, Jesse Harrington Au was there to guide me. He is a lifelong maker, but his story and way of teaching was a perfect fit for my learning style.

At fourteen years old, Jesse wasn't thinking about CAD. Like a lot of kids his age, he was thinking about skateboarding. In particular, he was wondering how he could build a halfpipe—a large u-shaped ramp—in his backyard in upstate New

York. His parents had finally given him the go ahead on the condition that he build the structure himself.

Before he could actually start building, though, he needed a plan. He wanted to visualize his creation, so he started designing his halfpipe in programs like 3DS Max and Adobe Illustrator, programs he had become familiar with through his interest in animation. Once he had the design, he was able to begin construction. He worried about how he would bend the plywood to fit the curved design he had made on the screen, but as soon as he started, he found the material to be as malleable as he needed. A few weeks and a hundred dollars later, Jesse had the halfpipe of his dreams. More important, he had been bitten by the design bug.

His projects grew in complexity, from designing solar heating devices with his dad and brother to creating complex art projects to fill large, empty rooms. Jesse was finding all sorts of uses for his burgeoning CAD skills. He was also developing a joy in sharing the maker skills he was learning. During college, he had a job working at a day care for kindergartners, and in an effort to get the kids building and making, he organized an activity for the kids to build their own guitars, drums, and horns. None of the parents or other daycare employees thought the kids would be able to hang in there with Jesse's plan, but by the end of it they had created their own band and were performing concerts for their parents. He also worked part time teaching CAD to other students at Rochester Institute of Technology. Jesse's diverse maker career eventually brought him out to San Francisco to work for the Exploratorium, building exhibits for the children's museum.

Jesse currently works as Autodesk's "Maker Advocate." He spends most of his time in makerspaces around the country, working with makers (especially new ones) to build their CAD skills and incorporate digital design into their maker toolbox. In the past few years alone, he's taught and assisted over a thousand new makers as they wade into digital fabrication. That's how we met. Jesse was teaching classes in Autodesk Inventor, Autodesk's product design CAD program, at Tech-Shop in San Francisco. The class is a complete crash-course, with Jesse explaining to me that he's squeezed the 15-week course he taught in college into a curriculum that teaches the material in three hours. I pressed him on how he was able to do it, and he attributed it to the eager curiosity of makers:

> *If you give passionate people the tools and time to play—learning just enough to play—then they'll get it every time.*

Here are some other tips from Jesse's experience:

Question: do you really need CAD?

CAD is the first step in the digital fabrication process, but it's important to remember that using CAD isn't always the right solution or option. It's great when you have something that's a very interesting or odd shape, like a curve that needs to be exact. It's also necessary for designs that you want to repeat or manufacture. But if you're rapid prototyping, and you just want to see if something will work, the best option may be just to use a table saw or another analog tool to quickly mock up the prototype.

Be patient; know your options

There are a lot of CAD programs out there for makers, even new makers. Over the past few years, a slew of new options have become available that are especially suited to new makers. It's worth trying a few of them to see which one feels the most comfortable:

Sketchup

It's free and easy to get started with. You might eventually find that you are limited in what you can actually build with it, and the file formats can be limiting.

TinkerCad

A great option for beginners (and also kids—more on this in Chapter 9). This browser-based program is incredibly easy to get started with, doesn't require you to download or install any software, and makes it simple to export to a 3D printer or printing service. It's great for primitive shapes, but trickier for more complex designs.

Autodesk 123D Design

The free 123D Design software is available for a number of different platforms: desktop software, browser-based, or even for your iPad. It is easy to get started with, but has some limitations in terms of more complex designs. The big benefit of learning 123D is that you can upgrade to Autodesk Inventor or Inventor Fusion 360 and have a very similar user experience.

What are you making? What's the output?

Are you making something mechanical? Or is just a replica of a character? If it's something that will move, the precision of CAD design can be critical because you need those exact measurements. If it's a model of your own head, CAD won't be as helpful. You'd be much better off using a laser scanning or photogrammetry device that can, essentially, take a 3D picture and translate it into digital bits.

Along the same lines, it also depends on your intended output for the digital design. Will it be 3D printed? Is it going to be used for manufacturing tooling? Thinking through the intended use will help identify the most effective mode of design.

Don't recreate the wheel!

Part of learning CAD is being aware of the vast libraries of designs that are freely available on the Internet. You don't need to create every part or model from scratch. Many of your designs can come from modifying an existing model, or assembling n set of parts. Here are some resources for finding existing designs:

Thingiverse (http://thingiverse.com)

This website was started by the MakerBot team to share the designs of desktop 3D printer users. Thingiverse has evolved into one of the largest repositories of 3D designs, where you'll find everything from iPhone cases to RC car wheels. Thingiverse tends to slant toward designs that work best for the desktop 3D printers.

Autodesk 123D Design (http://123dapp.com)

Similar to Thingiverse, Autodesk is also amassing a library of user designs for makers to use and modify.

GrabCAD (http://grabcad.com/)

GrabCAD boasts over 75,000 3D models, ranging from motor design to mechanical horses. The GrabCAD library is suited to a community of mechanical engineers, and the designs reflect a higher degree of sophistication and precision.

3DContentCentral (http://www.3dcontentcentral.com/)

Like GrabCAD, 3DContentCentral is a repository for thousands of predesigned files that you can use to build your model.

McMaster-Carr (http://www.mcmaster.com/)

McMaster-Carr is like the hardware store for the Internet. It has over 500,000 products: every screw, fastener, or fitting you could imagine. In addition, it has a CAD file for nearly every product. Know this before you try to redesign a screw or bolt!

Inventables (https://www.inventables.com/)
> In addition to the thousands of products and materials for designers, Inventables also has a collection of projects that include CAD files, a list of the materials required, video instructions, and a way to ask the designer a question. Most of the projects are made using tools including laser cutters, CNC mills, and 3D printers.

It's a language!
> Just like learning any other language, it's going to take some time. The only way to get better with CAD is to practice. Luckily, it's easier than ever to get started, and it will certainly be one of the most useful skills of the twenty-first century.

3D Scanning

Just as the CAD tools race to become cheaper and easier to use, 3D scanners and photogrammetry tools are also becoming more affordable (even free) and increasingly functional. These tools, like the MakerBot Digitizer or Autodesk's 123D Catch app, use lasers or a series of photographs to take what amounts to a 3D photograph of an object or scene, which can then be manipulated and modified as a digital design.

My first experience with 3D scanning came with the Next Engine 3D Scanning class at TechShop. The Next Engine uses lasers, a camera, and a rotating platform to scan objects into a digital model. At first, I was confused as to how and why this type of tool would be useful. It seemed like a novelty, only good for small trinkets. It wasn't until my classmate showed me a sea shell he had found and wanted to digitize, that I began to realize the usefulness of the tool. He explained to me that he had tried to re-create the shape in CAD, but wasn't able to do it. Scanning was the only way to duplicate the organic shape.

Since that class in the spring of 2012, a number of new maker products have become available, including the free iPad app 123D Catch. Catch doesn't require expensive lasers and platform, all it needs is the camera on the iPad (or any camera if you use the desktop software). With a series of photographs taken from multiple perspectives around an object, the app is able to model that object in digital form. I started where most everyone else does: trying to create a model of my head, as shown in Figure 6-3.

Figure 6-3. A model of my head

After seeing how easy it was, I immediately began to wonder if we could use this technology underwater with our OpenROV cameras. I had visions of using OpenROV videos to create 3D models of coral reefs, with the hope that easy-to-create 3D renderings would make it easier to visualize their health over time.

But first I had to see if the technique could work underwater. Unable to find any examples online, we decided to experiment with it ourselves. Zack and I ended up making a day-long excursion to the Aquarium of the Bay in San Francisco, pushing the boundaries of acceptable "touch pool" behavior by taking numerous photographs of starfish underwater. The results were surprisingly good.

Most important, that was our first attempt. When I asked the Autodesk team for other underwater photogrammetry examples, they didn't know of any. No one had tried it yet (with 123D Catch, anyway). *That's how new this stuff is.* Two guys with almost no experience, an underwater camera, and free software can test the limits of the shifting digital/physical boundaries.[2]

3D Printing

Ok, so you've got a digital design. What now?

One of the most exciting options is to print that design using a 3D printer. At a broad level, it's as easy as just clicking "print" and waiting for your design to appear.

Additive manufacturing, another name for 3D printing, isn't just one technology. There are actually several different processes that are classified as additive manufacturing. The most common technique for desktop variety 3D printers is fused deposition modeling (FDM), which pushes heated filament through an extruder down onto a build platform, creating a thin layer of plastic that can be built up.

---

2. Here's more on the starfish experiment (*http://bit.ly/14y6xS8*).

Other additive manufacturing technologies, stereolithography (SLA) and selective laser sintering (SLS), use lights or lasers to cure liquid resins or powder to form the desired 3D model.[3]

The promise of desktop 3D printing is that we'll be able to print many of our household items through one machine that sits on our desks. The reality, of course, is much different. It's not the panacea that some media articles have made it out to be, but it's also not just another toy. It's somewhere in the middle. Only time will tell exactly how people will integrate these machines into their lives. Regardless of its eventual impact, the magic of desktop 3D printing is an important part of the new maker experience.

3D printing isn't new—it's actually been around for decades. Inside university research labs and the R&D departments of multinational corporations, the technology has been used to create prototypes of concepts, products, and ideas. The recent 3D printing boom is less of a technological advancement, but more so an accessibility revolution. Expiring patents and fierce competition are creating a surplus of consumer options. And it's moving quickly.

At the World Maker Faire in 2012, there were over fifty different models of low-cost 3D printers, each one slightly different than the next. They vary in shape and size, cost and competency. Getting started with 3D printing has never been easier or more accessible, but it's also never been more confusing. Where do you start? How should you approach this world?

At the end of 2012, *Make:* magazine decided that question was worth answering. They decided the best way to answer that question was to get everyone in the same room and find out. They invited 15 of the leading desktop 3D printer makers to their offices in Northern California and put them through a weekend of rigorous testing and examination. They packaged the results of that weekend (along with essays from leading thinkers, resources for CAD and CAM, as well as a guide to 3D printing service companies) into a neatly designed "Ultimate Guide to 3D Printing" (it is available as a PDF in the Maker Shed (*http://www.makershed.com*)).

It's a beautiful and thorough resource, but it's quickly becoming outdated.

If you're interested in using a 3D printer or you want to build your own, the guide is worth your investment. Despite the (relatively) out-of-date information, it's still a good resource for thinking about how to evaluate different printers. If you're not sure what you would do with a 3D printer, I'd suggest playing around with one

---

3. Even SLA printing is finding its way onto the desktop, with Formlabs (*http://formlabs.com/*) raising nearly $3 million on Kickstarter for their Form 1 SLA printer in 2012.

first, either by using one at your local hackerspace or trying out one of the many 3D printing services that will ship your prints to you, such as Shapeways or Ponoko. Experiment before you invest.

Laser Cutter

"And that's just the beginning, there are all types of materials that work with the laser cutter. In addition to cardboard and paper, you can etch glass, cut acrylic, and engrave leather. You can even laser etch onto a chocolate bar," said Zack, my maker sherpa and TechShop Dream Coach. He was also today's Laser Cutting substitute teacher.

"Wait a second—chocolate, seriously?" I asked in disbelief. "Doesn't the laser melt it?"

"Nope, it works great. I used the laser cutter to engrave a picture and a poem on a chocolate bar last year for Valentine's Day. My girlfriend assumed I had organized a custom mold at the Ghirardelli factory, but it was only five minutes of laser cutting time after work one day." Zack continued, "I actually have a bar of chocolate in the freezer, we can try it out right now."

Next thing you know, everyone in the course was eating a piece of chocolate that they had just laser-cut their name into.

One of my main assumptions starting out on this journey was that I wouldn't be able to make anything cool right away. I thought I'd be exposed to different tools and processes, but it would take years of practice and many mistakes before I could do anything useful. The Laser Cutting course at TechShop, shown in Figure 6-4, totally blew that assumption out of the water.

It started off just how you would expect: basic safety information and an overview of the machine (on/off, cleaning the lens, orientation, etc.), but Zack quickly let us loose to try both raster cutting (used for engraving/etching) and vector cutting (used for clean cuts through material) for ourselves. It was such an easy process to learn that I was a little embarrassed that I hadn't attempted it sooner. I think I harbored a bizarre fear that I needed more CAD experience or some other technical know-how, which is not the case at all. The machine runs from Adobe Illustrator or Corel Draw. It's as easy as typing, drawing, or uploading an image you want to use and sending a print job (with a few settings tweaks) to the laser cutter.

Figure 6-4. The laser cutter course

Aside from the general ease of use, I was surprised at how many ideas for using the laser cutter were pouring into my head. I could use it to personalize my wallet, create a mold for a ceramics project I was envisioning, or repeat Zack's chocolate etching. The laser cutter experience was also valuable for our OpenROV project. We had recently redesigned the frame of the ROV to be cut from a single sheet (24 × 18 inches) of 1/4 inch-thick acrylic. After a quick heat bend of the main section, the rest of the pieces snapped into place. Using a single sheet (as opposed to multiple materials, connected with adhesives and fasteners) cuts the cost of the ROV dramatically (which is the main goal of the project) and makes it fast and easy to reproduce (the other main goal). After taking a two-hour course, I could easily contribute to creating OpenROV structures.

I'm not alone in my admiration and appreciation for the laser cutter. In *Makers*, Chris Anderson called the laser cutter "the real workhorse of the Maker Movement" because it's the one that everyone uses first and most often. TechShop staff (and others) refer to it as the "gateway drug" to digital fabrication. It's how Bilal initially became involved with the maker movement, breaking into the local art school to create stencils for his printing business. After you make something with a laser cutter, it's hard to go back.

Getting Creative with 2D Design

There are several advantages to using the 2D digital fabrication tools, like laser cutters, in lieu of the sexier 3D printing methods. Most notably, the 2D fab tools are much faster. If you plan to do any kind of small batch production run—from 20 to 1,000—you'll care *a lot* about speed. It's possible to use a laser cutter for this type of micro-manufacturing (when you get into the 1,000+ range of production runs, you'll probably need to develop a broader manufacturing strategy). To give you an example of the time savings, laser cutting an entire OpenROV shell (the acrylic structure) takes less than 20 minutes. In contrast, it would take almost a full day on a 3D printer. And, at this point, the desktop 3D printers aren't fit for production tolerances. With the laser cutters, we get precision cuts every time—good enough for our OpenROV kits. With desktop 3D printers, it's not nearly as consistent.

However, don't let 2D tools limit you to 2D designs. There are a number of strategies to extend the utility of the laser cutters and CNC mill machines into the third dimension:

Origami

I'm continually blown away by the creative ways I see makers folding and bending their designs into more useful shapes. We spent a lot of time trying to arrange perpendicular pieces of acrylic for the outer shell of our OpenROV. One day, while we were working at TechShop, another member suggested we use the strip heater to bend one piece of acrylic instead of trying to attach two pieces. Eric and I looked at each other with bewilderment. We had never even seen the strip heater in the corner of the shop, and had never thought about bending our design. It worked perfectly. It was cheaper, easier, looked better, and gave the ROV more structural integrity.

I've seen an entire origami kayak cut from one piece of corrugated plastic board (more on this in Chapter 7). With a little imagination, a series of bends and folds can go a long way.

Stacking

Anything that can't be accomplished with folds can likely be figured out through assembly. You don't need to be an experienced mechanical engineer or designer to use this strategy. Autodesk 123D Make is a software service that automatically dissects a 3D model into 2D shapes marked for assembly, and ready for the CNC machine of your choice.

Note

Pro tip: Use 123D Make, cardboard, and Bondo for a quick and easy way to create a mold. Check out this guitar mold used for laying fiberglass.

Boxes and hinges

The craft of joinery is nothing new, especially among the Japanese. Even though Wikipedia defines the terms "joiner" and "joinery" as being obsolete in the United States, the lessons of attaching wood without nails and screws is having quite a revival in the form of augmenting CNC'ed or laser-cut structures. As many makers are learning, clever arrangements can go a long way.

In fact, the entire internal structure of OpenROV is actually all 1/4 inch acrylic plastic elegantly snapped together.

Make: magazine writer Sean Michael Ragan has assembled an exhaustive resource (*http://bit.ly/166hmdQ*) that points out some of his favorite methods, shown in Figure 6-5.

Note

Pro tip: Makers always seem to need to design boxes. We've needed them for the OpenROV topside adapter as well as a dozen other things. I've seen many people need them for housing electronics, and others for gardening kits. For some reason, this *always* comes up. Making all of the notches and dimensions fit together can actually be a big design headache. Luckily, BoxMaker (*http://boxmaker.rahulbotics.com/*) is a service that automatically creates box designs around your parameters.

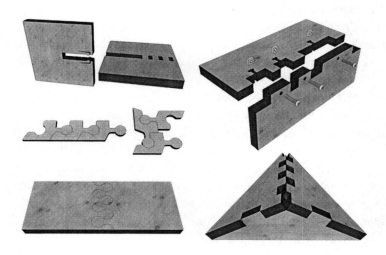

Figure 6-5. Photo by Sean Michael Ragan

Digital Fabrication Is Not Just 3D Printing

3D printing grabs all the headlines, but as many makers will tell you, 3D printing is actually a small part of the digital fabrication revolution. When Neil Gershenfeld first started the fab lab project, he described a list of tools—laser cutters, CNC mills, and vinyl cutters, tools to program microcontrollers—that could create (almost) anything but also completely replace themselves. 3D printers weren't even a part of that original tool list. To reiterate his initial hypothesis and to address the growing media attention given to 3D printing, Gershenfeld commented, at the Science of Digital Fabrication conference at MIT in 2013 (*http://bit.ly/17H6h5I*), on the recent 3D printing hype as "like telling chefs in the 1960s that microwave ovens are the future of cooking."

It's a very limited view of what's actually happening.

Admittedly, I didn't grasp the true potential of digital fabrication until I was several months into the journey. The 3D printer caught my attention with its novelty. The laser cutter surprised me with its ease of use. But it wasn't until I first used a CNC mill machine during a ShopBot class at TechShop that I truly grasped what was going on.

The ShopBot class was a long time coming. In addition to the hands-on classes and experiences, I was also learning the software side of things. As part of the membership package, TechShop offered a list of CAD classes based on Autodesk software. But there were also CNC courses that take place in the computer lab, such as one on CAD to CAM, the software used to translate the digital designs into the programming language for CNC machines and G-Code. Even though I didn't have a clue what those letters meant, or what I was going to use it for, everyone kept reiterating their importance. They were right.

Something clicked during that ShopBot class. I realized how, with some combinatorial creativity, you could take just about any idea from digital bits on a computer screen to something real and tangible. Whether it was 3D printing, laser cutting, or CNC milling, there was almost always a way to bring those digital files to life (and then to share them). More important, I realized how accessible all these tools actually were. I mean, I was actually using them! Sure, I had been taking classes for the past few months. But that was it. No engineering degree. No garage tinkering childhood. Nothing. But here I was, with the power of a micro-manufacturing operation at the tips of my fingers.

Desktop CNC Mill

When I began writing this book, I wasn't sure how to talk about computer numerical control (CNC) milling or machining. It's one thing to understand the description: computers that control machines, allowing for more consistent, repeatable cuts and shapes.

But for me, sitting in a class and hearing someone explain it was nowhere near the experience of using the ShopBot for the first time. Writing about it, I assumed, was going to be nearly impossible.

I also assumed that it wasn't as accessible as some of the other digital fabrication tools, like 3D printers and laser cutters. TechShop has several ShopBot and CNC Tormachs, but most hackerspaces are lucky to have a MakerBot, or really lucky to have a laser cutter. I had seen some of the open-source, DIY attempts, such as the MyDIYCNC, but they just didn't come close to the functionality I was getting from the ShopBot.

Then, my personal maker path took me off in my own direction—laser-cutting acrylic for our robots and programming Arduino boards to control our OpenROV motors. Out of preoccupation, I pretty much stopped thinking about desktop CNC machining. It wasn't until a conversation with Zach Kaplan, the CEO of Inventables, many months later that I realized how much I was missing out on in the desktop CNC department.

Zach told me the story of ShapeOko; a story I *thought* I knew. In the summer of 2011, Edward Ford launched the ShapeOko project on Kickstarter, with a funding goal of just $1,500 and a humble mission to create an open-source CNC design that anyone could use to build a simple CNC machine for less than $300. He didn't even have a video for his Kickstarter project, just a photo of a laser-cut plywood CNC machine. It didn't have a mill head or Dremel tool like other DIY attempts. The original photo, shown in Figure 6-6, just held a pen.

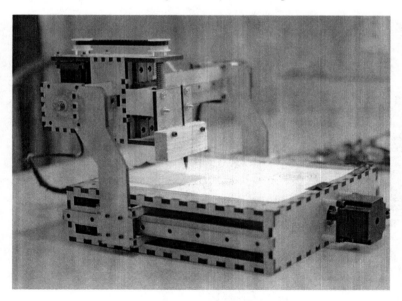

Figure 6-6. Edward Ford's photo of the original ShapeOko

Edward ended up raising more than $11,000 from 123 backers during the month-long campaign. Respectable when you consider his original goal, but far from a raging Kickstarter success. He didn't have plans to sell kits afterward, either. He thought he'd just release the designs and that would be it. And that's about where I stopped paying attention.

But when I talked to Zach, I was surprised to hear that he had actually hired Edward to work at Inventables, upgraded the ShapeOko with new materials, and was now selling kits of the machine through Inventables. And it wasn't just a few machines, either. In its first year of sales through Inventables, they sold more than twice the number of 3D printers that MakerBot did in its first year.

It's no longer the laser-cut, pen-holding machine that Edward originally put on Kickstarter. The ShapeOko has gone through serious upgrades as seen in Figure 6-7. Instead of the lasercut wood, it now uses Makerslide (another Kickstarter-funded creation) beams as the functional base of the machine. The Makerslide upgrade enables the machine to be expanded to a cutting platform of 6 × 6 feet. That's almost as big as the largest ShopBot in TechShop!

Figure 6-7. The evolved ShapeOko (photo by Inventables)

With ShapeOko kits coming in at $599, I realized that even CNC milling is coming to your desktop (or your garage, or at the very least, your local makerspace).

CAM (CNC Software)

Zach and Edward didn't stop at creating an affordable desktop machine. They knew that if they really wanted to make CNC machining an accessible option for makers and hobbyists, they would need to make the software side of the equation just as approachable.

Computer-Aided Manufacturing (CAM) software is like a bridge. It's what translates the digital designs from their native file formats like *.dxf* or *.stl* into a language that the CNC machine can understand. More specifically, it needs to understand what to cut (like the actual shape of the object) as well as how to cut it. The *how* to cut information includes instructions for how fast the endmill should spin, how deep it should make each pass, what order it should make the cuts, etc. These are the "feeds and speeds," as they're called. There are a lot of rules that govern how best to optimize. I've barely scratched the surface myself.

To streamline that process, Zach and Edward have adopted MakerCAM, an open-source, web-based CAD program created by Jack Quio. The program helps you easily create simple shapes and designs and export them to your CNC machine. It's an easy-to-use drawing program with added feature options such as the ability to control the target depth and plunge rate (both "feeds and speeds" terms).

They've set up a tutorial (*http://www.makercam.com/tutorial.html*) that also serves as a great introduction to CAM and CAD.

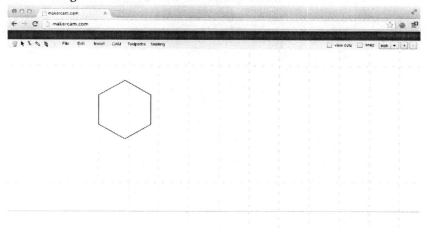

If and when you continue your CNC practice, you'll eventually want to move to more robust CAM software like VCarve Pro or CUT3D.

Electronics 101

Do you know Ohm's Law?

I didn't. I mean, I'm sure I learned it at some point in my life, but I couldn't remember when the instructor of the Basic Electronics class at TechShop looked around the class and asked us all to confirm our understanding. He nodded at the five of us, with an "of course you know what Ohm's Law is" look on his face. I looked at everyone else and they replied with a similar nod. Uh oh.

I quickly whipped out my phone and googled "Ohm's Law" so I wasn't completely in the dark:

$I = V/R$

It means the current (I) is equal to the voltage (V) divided by the resistance (R). For someone like me who has never tinkered with electronics or taken an electrical engineering course, I'm not sure there is a good reason to know this, but I still felt stupid for not knowing. I also worried that I might have hit a major snag in my quest to make things. How much electrical engineering was my education lacking? Would I have to go back to school for this?

I sat through that class, soaking up as much as possible. Some of the talk about current and resistance went over my head, but some of it made sense. It was the tangible, hands-on part of the class that really helped me understand. This was one of the first times I had used a soldering iron, and I was still getting used to the feel of it in my hands. The beginning activities of soldering together paperclips and basic breadboarding gave me a tactical sense of how electricity was moving through these systems and how I would be able to manipulate it.

It was the actual experience of electronics that helped me to understand it, not the book reading or theoretical explanations. This has remained true for my entire electronics education. I've fried boards, blown LEDs, and shocked myself. I've made lots of mistakes, and I'm still learning *a lot*. Nowhere does my "enough to be dangerous" theory apply more than with electronics. I'm roughly aware of how much I don't know and have developed a mental framework for working through challenges.

It's worth mentioning how important understanding electronics is to making. It's certainly possible to take part in the maker movement without learning any electronics. Joel Bukiewicz, the knife maker, is a perfect example. However, many of the most interesting projects and opportunities require at least a basic level of electronics understanding.

You don't have to learn *everything*, though. So far, I've only learned as much as the OpenROV project has required of me: powering an Arduino and BeagleBone, powering motors, electrical shorting in water, battery chemistries (and how they relate to buoyancy). That's a very small sampling of the entirety of electronics, but it's gotten me where I need to be.

Framing your electronics education in terms of a project like OpenROV is a great way to make the learning relevant and manageable in scope. Now, if I wanted to start another electronics project, I have the base knowledge of everything I learned through building OpenROV to build off.

This project-based, experiential style has worked wonders for me and has turned electrical engineering from an intimidating academic study into an interesting series of experiments.

Charles Platt has written an entire book on the subject for *Make*: based on this style called *Make: Electronics* (*http://oreil.ly/13RQoqT*). I started with this book and highly recommend it to other beginners who are interested in building a working knowledge of electronics for their future projects. If nothing else, it will certainly do a better job explaining Ohm's Law.

Arduino and Beyond

As broad and intimidating as learning electronics can be, it was surprising to me how quickly I was able to actually start *doing stuff* with what I was learning. And I mean more than just blinking an LED light. A basic understanding of electronics is enough to get you working with Arduino, "an open-source electronics prototyping platform based on flexible, easy-to-use hardware and software."

The Arduino boards are the gateway to making your projects move, listen, sense, and react. The board uses a microcontroller and very user-friendly programming environment to bring your designs and projects to life.

Unfortunately, my simple definition doesn't do Arduino justice. Phil Torrone (*http://bit.ly/1eZGx7h*) came up with the better explanation:

The "what" is still a little vague, and that's the Arduino's strength. It's the glue people use to connect tasks together. The best way to describe an Arduino is with a few examples.

- *Want to have a coffee pot tweet when the coffee is ready? Arduino.*
- *Want to have plushie steaks glow? Arduino.*
- *How about getting an alert on your phone when there's physical mail in your mailbox? Arduino.*
- *Want to have a Professor X Steampunk wheelchair that speaks and dispenses booze? Arduino.*
- *Want to make a set of quiz buzzers for an event out of Staples Easy Buttons? Arduino.*
- *Want to make a light-up arm cannon from Metroid for your son? Arduino.*
- *Want to make your own heart rate monitor for cycling that logs to a memory card? Arduino.*
- *Want to make a robot that draws on the ground, or rides around in the snow? Arduino.*

Don't let the programming aspect scare you off. As someone who had never worked with electronics or learned much more than basic HTML and CSS programming, I was worried. But I quickly got the hang of it once I understood what the microcontroller was capable of: listening to commands (like a key on your keyboard) or sensor input (like gyroscopes, accelerometers, or compasses) and turning that into physical interactions like controlling actuators (LEDs, servos, motors, or electronic speed controllers). I'm by no means a programmer, but I've been able to effectively use the intuitive Arduino developer environment. In fact, I'd say that learning to program Arduino boards has actually increased my programming abilities and understanding beyond microcontrollers—a perfect stepping stone into programming.

And, like everything else in the maker world, there are broad shoulders to stand on. The Arduino community has created vast libraries of code to do everything from read accelerometer data to control electronic speed controllers. The community has done almost everything imaginably possible with a microcontroller. If not exactly, there is usually a chunk of code that is at least similar to your specific need. For many projects, getting the Arduino to behave the way you want it to is a matter of finding the right library and modifying it to your needs.

Where the Arduino leaves off, makers have picked right up. The falling costs and ease of manufacturing of printed circuit boards (another Gershenfeld vision for the future) and the open-source modularity of the Arduino have created a Cambrian explosion of "shields," additional boards that can be plugged on top of the Arduino to extend its capabilities. A quick tour of Maker Shed, Adafruit, or Sparkfun will turn up dozens of options for shields—everything from LED displays to GPS loggers, as well as "protoshields" that allow you to design your own custom shields.

Some combination of intuitive design, ease of use for non-engineers, extensive libraries, open-source development, and a large community of supportive developers have made Arduino the go-to prototyping platform for maker electronics projects. Phil Torrone was right: for makers, Arduino is here to stay.

Here are some more resources:

- Arduino page (*http://www.arduino.cc/*)
- Make: Arduino page (*http://makezine.com/arduino/*)
- *Getting Started with Arduino* (*http://bit.ly/14yaPZM*) by Massimo Banzi (co-creator of Arduino)
- Sparkfun Arduino Guide (*https://www.sparkfun.com/pages/arduino_guide*)
- Adafruit Learning System (*http://learn.adafruit.com/category/learn-arduino*)

Microcontrollers aren't the only computing devices participating in the maker movement. Now, microprocessors, the "thinking" computers that run the devices like your smartphone or tablet, are getting involved. Miniature Linux computers, or "systems on a chip" (SoC), like the Raspberry Pi and the BeagleBone, are becoming important maker tools, and adding another complete layer of functionality to project possibilities.

Raspberry Pi, a $35 general purpose computer, created waves when it was announced, and eager customers waited for months to get their hands on one. It's been a few years since the release, and there are now millions of these devices in the wild.

The difference between Arduino and Raspberry Pi or BeagleBone is an important one. The Raspberry Pi doesn't replace the Arduino. It actually adds to it. Take our OpenROV for example, which runs both a BeagleBone as well as a microcontroller board inside the small underwater robot. The BeagleBone does much of the heavy lifting, like processing the digital video from the webcam and running the Node.js software that is used to access control of the ROV from your browser. However, the BeagleBone is not great at running motors, so we've added a microcontroller to our BeagleBone *Cape* (same concept as an Arduino Shield—an additional PCB with additional functionality) that runs Arduino code in order to control the motors and servos. The OpenROV blends the best of both worlds.

These new miniature computers are so new that we're still learning what's possible with these devices. Already, we're starting to see robots with computer vision and arcade game coffee tables. Especially if you're already familiar with Linux and software development, the opportunity to create new and useful projects with these tools is wide open. It's a whole new world!

Here are some more resources for you to check out:

- *Getting Started with BeagleBone (http://oreil.ly/19GCP5y)* by Matt Richardson
- *Beaglebone Quick-Start Guide (http://bit.ly/18HFAPs)*
- *Getting Started with Raspberry Pi (http://oreil.ly/1dpdFr4)* by Matt Richardson and Shawn Wallace
- *Raspberry Pi Quick-Start Guide (http://www.raspberrypi.org/quick-start-guide)*

One to One Thousand

Today, digital fabrication tools are more than just accessible. They're also powerful.

I didn't notice it during the initial classes I took. At the time, it just seemed like learning; everything was new and interesting and overwhelming. The first three hours of laser cutting looks a lot like the first three hours of any manual skill, like wood carving or welding. It's a period of constant absorption and productive struggling.

But that's not where the digital fabrication tools shine. Having their design DNA embedded into bits means that it's cheap and easy to create an exact copy. Getting to the prototype or mock-up stage always takes effort, whether digital or analog. With a digital design, though, you don't have to re-invent the wheel for the second revision. The leverage isn't going from zero to one, it's going from one to one hundred (or even several thousand).

Although it requires more up-front design time, the amplified production capacity afforded by the computer-controlled machines means that making many (anywhere from hundreds to thousands) of one design is relatively straightforward. This is beyond prototyping. As many new makers are finding, this micro-manufacturing can lead to some interesting and exciting business opportunities. The challenge, as with all business, is finding a market for those products. But makers are learning that, too. In the same DIT style used to learn the new tools, we're helping each other to turn these projects into burgeoning enterprises.

Perhaps nobody has lived this Maker Dream more authentically than Abe and Lisa Fetterman.

Sous Vide Dreams

Abe is tall, thin, soft spoken, and whip smart. He has a degree in physics from Caltech and a PhD in astrophysics from Princeton University. His laid back attitude and kind demeanor make him approachable and engaging. He's shy and humble and intellectually intimidating. A conversation with Abe leaves you more confident in the human race but sharply reminded of how little you know.

Lisa is the most outgoing person I've ever met. She's Chinese-American, with a pair of thick-framed glasses and a huge smile. She has that rarest of abilities: she makes you feel both uncomfortable and comfortable at the same time. Her filterless commentary keeps you on edge, but her smile reassures you that there's a heart full of kindness behind the gregarious shell.

If you met them separately, you'd never pick them out as a match because of their wildly different personalities. When you meet them together though, you wonder how they ever lived apart. They're a dynamic couple, and their attraction is magnetic.

I first met Lisa and Abe at the World Maker Faire in New York City. They were our neighbors for the weekend, with our booth for OpenROV adjacent to their Arduino-powered, DIY sous vide cooking device. They drew a much bigger crowd than our project, mostly because they were serving up deep-fried egg yolks to anyone who passed by. Of course, people were interested in the fried egg yolk, but more so, they were attracted to the energy and passion of Lisa and Abe.

Over the course of the Faire weekend we shared a few pleasant conversations, but we were both so busy with fairegoers that we didn't have time for much of an extended conversation. It wasn't until a year later, while on a story assignment for *Make*: to cover the Haxlr8r demo night in San Francisco, that I would again cross paths with Lisa and Abe. Only this time, the circumstances were very different.

The demo night was the culmination of the Haxlr8r program, a 15-week boot camp for budding hardware entrepreneurs that took them from the factories in Shenzhen, China, to the venture capital offices of Silicon Valley. The program was the first of its kind: a start-up accelerator that took advantage of the new rapid-prototyping environment to help aspiring entrepreneurs make their products along with a properly organized business. I sat in the front row, jotting down notes for the blog post. I found all the pitches and demos to be of good quality, certainly more interesting than the countless number of app-making or Facebook-clone startups you run into around San Francisco. When Lisa took the stage as the final presentation, my heart fluttered with excitement.

To say I was shocked would be an understatement. It wasn't just that I was surprised to see them, but to see how far they had come in only one year was hard to wrap my head around. They had turned their hacked-together DIY kit into a beautifully designed kitchen product—something you could envision buying at Target or Bed Bath & Beyond. As it turns out, the full story of how they went from DIY amateurs to Food Network-ready entrepreneurs completely lived up to my demo-night amazement.

Two years earlier...

Lisa and Abe were both living in New York. Abe was working as an astrophysicist and Lisa was studying journalism at NYU. They had been dating for about a week when Lisa made an offhand comment about wishing she could cook *sous vide* in her apartment. Sous vide is a method that involves slow cooking food inside of a plastic bag in a water bath at a precise temperature over long periods of time. It was popular among the high-end chefs that Lisa admired, but the home-use machines were prohibitively expensive. Abe, ever the enterprising swooner, promised he would make her one. He didn't have any experience in this type of endeavor, but possessed plenty of confidence nonetheless.

At this point in the story, neither of them knew how to solder.

Their lack of relevant education and basic making skills didn't slow them down one bit. With a reckless but admirable confidence, they threw themselves into the project of creating their own sous vide machine. It wasn't long before they had created a makeshift device. It consisted of only $50 in off-the-shelf parts (and involved no soldering). They published their design on their blog (*http://qandabe.com*).

"We thought it was going to blow up the Internet, but nobody came," Abe confessed to me. Their design, even though it was the lowest-cost DIY sous vide around, didn't get much attention.

They kept building. Soon, they had created an improved DIY model that could be built for $75 in parts. But, again, not much attention.

It wasn't until a chance encounter in a Manhattan coffee shop with Mitch Altman, all-around maker superhero and creator of the TV-B-Gone, that their story took a turn for the wonderful. At the time, Abe and Lisa didn't know about *Make:*, Maker Faire, or who Mitch was. They were just sitting at a nearby table as Mitch was being interviewed by Matt Mets. They overheard the entire interview, and after it was over, they approached Mitch.

"Hey! We're makers! I think..." Lisa said to him.

Mitch invited them to a soldering class he was teaching at Alpha One Labs and Lisa took him up on it. With the knowledge she learned there, and some Arduino skills they picked up at NYC Resistor, they designed a DIY sous vide kit called the "Ember" and began selling it for $80.

I stopped them at this point in their story. I was doing the math in my head, "Wait, that must have been a slim margin on your kits. Right?"

"Oh, certainly. There was no margin at all. We just wanted everyone to be able to sous vide," Lisa told me. And they did. They made their kit as easy to assemble and use as possible. If it was something they could figure out, they thought, then anyone could do it. They started teaching classes at Alpha One and NYC Resistor to anyone who was interested. While teaching one of their classes, they met a native Thai chef who was working in the city named Bam Suppipat (he comes back up later in the story).

Pretty soon, life got in the way of their sous vide dreams. Abe got a job in San Francisco and the couple relocated to the Bay Area. Lisa also began working a new job. Their passion for cooking seemed as though it would always remain a hobby.

But they missed it. After seeing a write-up about the Haxlr8r program, Lisa and Abe decided to go all-in on their sous vide idea and try to turn their dream into a business. They were committed and now they had no other options but to try to make it happen.

Once they were in Shenzhen, China, everything became harder. They were quickly running through their seed capital and had very little to show for it. Out of ideas and stressed about their project, they decided to take a short trip to Thailand to clear their heads. They called the only friend they had there, Bam, and asked him to show them around.

During their first night with Bam, they explained their sous vide project and all the challenges they were facing, technically and emotionally. As it turned out, he was the perfect person for them to confide in. In addition to his studies at the French Culinary Institute, unbeknownst to Abe and Lisa, Bam also had an Industrial Design degree from the Rhode Island School of Design. His life's dream was to design better culinary devices and equipment, but he had recently resigned himself to moving back home to Thailand and taking a stable corporate job. Even though he was stuck in a cubicle during the week, Bam was still cooking and teaching low temp cooking in Thailand on weekends to home cooks.

It was a match made in heaven.

What started as a friendly evening of catching up quickly turned into a full-blown design intervention. The team spent the next three days reviewing, designing, and imagining what would eventually become the current Nomiku design. Abe and Lisa headed back to Shenzhen with a renewed sense of determination, and Bam, who still couldn't believe that his dream job had fallen into his lap, joined them.

The team spent the next month building, developing, and sourcing the design that became the Nomiku Sous Vide Cooker. By the time the Haxlr8r program came to an end, they were ready to take the next logical step for any maker business: put their project on Kickstarter.

When I saw them at the demo night they were closing in on having raised $100,000 of their $200,000 funding goal. By the time the month-long Kickstarter campaign had ended, they had garnered more than $586,000 from over 1,800 backers.

What started as an offhanded comment from Lisa to Abe had turned into a fast-growing company. They, quite literally, created a multimillion dollar business from off-the-shelf parts and the courage to follow their passion.

Define Your Own Success

Of course, Lisa and Abe's story is uniquely theirs, just as your maker story will be uniquely yours.

The goal of re-skilling yourself doesn't need to mean being an entrepreneur or building a business like Lisa and Abe. Re-skilling can be a powerful and effective way to find a new job or advance your career. It's also perfectly wonderful if making remains a hobby, something done on the side for fulfillment and enjoyment. Or maybe it's something you're doing to help arm your kids with skills for the future. It's all up to you. The best part is that you don't need to have a plan when you get started.

Regardless of whether making turns into a fast-growing startup, a small life-style business, or an enjoyable hobby, it all looks the same for new makers—makers like me, Lisa and Abe, or you. It starts as an exploration and, as such, requires a mind that is open to new ideas, new people, and new possibilities.

If you're looking to make a business or career out of being a maker, here are some important things to consider:

Make something that excites you

I know this advice is played out, but I mean it. All the interesting and successful makers I know, especially the new makers who have picked up rapid proto-typing skills in the shortest amount of time, are wildly passionate about and interested in whatever they're making. Lisa Fetterman loves food and cooking. Alex Andon was a marine biologist and loves jellyfish. Eric and I are excited about oceans and exploration.

The opportunity for digital fabrication tools and online communities to re-imagine and redesign the built and "made" world is ripe. It's a big, open field waiting for maker entrepreneurs to fill in the gaps. As you look to stake your claim, why not seize the moment and do something you've always dreamed of? The passion is the secret sauce, and it's what attracts the help, from the local maker scene as well as the larger global community of interest. In the end, it's the size and enthusiasm of the community that determines if the effort is worth turning the corner and becoming a business.

Also, I'd be remiss if I didn't mention that starting a business is *hard*. No matter how passionate you are, or how easy the tools are to learn, or how low the barrier of entry to distributing and selling your product, it's still a long, uphill slog. It's *always* harder than it looks. A maker business is no different than a regular business in that respect. It's only manageable if you're passionate about what you're building.

Decide what matters

In the era of micro-manufacturing and small-batch production, the goals don't necessarily need to be about making gobs of money and taking a company public. In fact, that's my favorite part about the next industrial revolution. It's less about enriching and enabling a select few manufacturers to become large, and more about allowing a larger number of niche manufacturers to remain small. The economics of being a small business finally work for the manufacturing sector. In *Makers*, Chris Anderson noted the following:

> What's interesting is that such hyperspecialization is not necessarily a profit-maximizing strategy. Instead, it is better seen as meaning-maximizing.

Eric and I have always been clear that the OpenROV project, even though we've become a company and are selling kits, was never about maximizing profit. We're trying to maximize our *Return on Adventure*; we wanted the project to add something to our lives and the lives of everyone in the OpenROV community. The strategy has paid off. Of course, we're leaving a lot of money on the table by selling the kits at the lowest price possible, but we'd rather more people have access to the ROVs. We're probably leaving ourselves open to cloners by posting our design files, but it's allowed us to meet and collaborate with like-minded people all over the world. We've been invited to dive with NASA at their underwater reef base off the coast of Key Largo as well as join scientists as they study whale sharks off the coast of Mexico. We might never become millionaires because of OpenROV, but we'll certainly have the ride of our lives.

Your maker business, like your maker journey in general, can be uniquely yours. The new playing field gives you the opportunity to maximize your own meaning, whether that's time spent with family, a community of collaborators, or the freedom to travel. Whatever it may be, the best chance at achieving it comes from being intentional about it up front.

The new maker economy is about more than just making your own product or business; it's about making your own meaning!

The Right Way to Use Kickstarter

As the saying goes, you don't know what you have until it's gone. This is a truism that Anton Willis faced when he decided to move to San Francisco. After growing up in Mendocino County in northern California, Willis was used to the wide-open spaces and outdoor recreational opportunities afforded to the rural lifestyle. This access to the outdoors, particularly to lakes, rivers, and the ocean, is what first fostered his interest in kayaking.

After he moved to San Francisco, however, his long fiberglass kayak didn't quite work. He didn't lack access to great kayaking space—there was ample opportunity in the San Francisco Bay. The real problem was that Willis' small studio apartment didn't have the space in which to store something that large. He was forced to pack the watercraft into storage and dream about another way to enjoy the water.

Months later, Willis' daydreaming found an unexpected source of inspiration while he was reading a profile of the origami physicist Robert Lang in *The New Yorker*. Willis wondered if he could apply the origami principles to his conundrum of needing a kayak that took up less space.

He set to work testing his hypothesis. He began with paper, folding up different design ideas until he had something worth prototyping on a larger scale. He worked his way through several dozen iterations on his way to a working prototype that he was ready to show the world: Oru Kayak (*http://www.orukayak.com/*), likely the world's first origami kayak.

Apparently, Willis' desire to have a functional kayak—one he could pack neatly into a suitcase-sized package and quickly unfold for use—was a more universal need. He put the project on Kickstarter and raised over $100,000 on the very first day. The month-long campaign netted over $443,806, with 730 backers chipping in to support.

It's the new American Dream: a person has a wild and creative idea, prototypes it until he gets it right, throws the project up on Kickstarter, makes gobs of money, and starts a new business. On the surface, that's easy to digest. It's the story all the blogs and magazines want you to believe. However, for new makers, this oversimplistic idea can be dangerous.

I've developed another perspective for thinking about the maker process, from idea to prototype, to product and business. I call it the maker-to-audience ratio (M:A ratio).

Admittedly, I didn't come up with the term. I first came across this concept while perusing photos by Dan Parham, who had a series from a local Legong dance performance in Ubud, Bali. Although the photos were visually stunning, I was more impressed with Dan's observation that there were 15 percussionists and 10 dancers entertaining a crowd of 10 people. An artist-to-audience (A:A) ratio, he noted, of 5:2.

My first thought was that the planning, preparation, and execution of a performance with an A:A ratio higher than one required an entirely different perspective on audience engagement. I imagined the Legong performers fixating on exactly how each member would perceive the performance. The more I thought about it, the more I realized that the A:A ratio could be an important lens for all artists and makers to consider. The applications (and interesting milestones) spread far and wide in the realm of artistic creation as well as entrepreneurial endeavors, especially for makers.

IT ALWAYS STARTS WITH ONE

No matter how experienced a maker you are, every new project starts out with an audience of one: yourself.

In the past year, as I've explained my re-skilling story to other reluctant makers across the country, there's a consistent excuse (one that I myself harbored for a long time) that seems to be holding a lot of people back: they don't know what they want to make.

It's profoundly common and perfectly reasonable. I mean, why would this newfound access to tools and expertise get anyone excited if they aren't sure what they would use it for? It makes sense.

Most people jump to the conclusion that they don't have any ideas for building something that *other* people would want. They don't believe that they have a useful innovation or design that could be made and sold. When all of the Kickstarter success stories and maker businesses in the news talk of overwhelming product demand, it's easy to be intimidated into thinking that that's the definition of "making it."

Don't be fooled. All of the great maker stories (and correspondingly great maker businesses) that I've encountered have come from people or teams that made something that they themselves actually wanted.

Anton made the Oru Kayak because he couldn't fit his kayak into his apartment. Lisa and Abe created their first DIY sous vide machine because they wanted to use the cooking technique but couldn't afford the expensive commercial models that their favorite chefs were using. Alex Andon wanted his own desktop jellyfish tank. Eric and I never planned to be selling ROVs, it was always something that we wanted for ourselves, so we could do our own exploring of the Hall City Cave. Making something you want is always the first step.

Over and over again, I hear stories of makers creating successful products and businesses as outgrowths of something that they personally wanted. Very rarely do I hear about a business that started from someone who set out to create something he thought had a very large market. If anything, that's the ballgame for GE, Walmart, and other large manufacturing companies who have money to throw at focus groups and marketing campaigns as well as the production and distribution channels to deliver those high quantities.

As an individual, it's almost impossible to compete at that level. And, personally, I wouldn't want to. Playing in the long tail end of the manufacturing scale allows for much more creativity. It allows commerce to happen at a more human scale. The trick to the new maker economy is finding your own unique niche. The best way to do that is to make something that you actually want.

It sounds easy enough: all you need to do is make the one thing you really want. But most people struggle with the magnitude of the question. In an op-ed piece for CNN (*http://bit.ly/13OL4Vb*), Jim Newton, Founder of TechShop, talks about this seemingly simple question:

> When I meet people who are not yet members of TechShop, I like to ask a simple question: "What do you want to make?
>
> The typical response is "Oh, nothing. I wouldn't know where to start," or "I'm not handy ... I don't know which end of a hammer to hold."
>
> Then I'll press further. "Isn't there something that you've wanted to make that doesn't exist? For your house or car, a gift for someone, or to improve your life or someone else's?"
>
> That's when an interesting thing happens. They light up and say something like, "Well, there is this one idea I have." They will describe the idea in great detail right down to color, variations, and the brand name they have given it.

I've had a similar experience. I've found that that thing—the creation that a person really wants to make—is a few questions deep. It takes a little digging to get it out. As Jim mentions, there's a light in their eyes that gives it away. Sometimes it's something they've always wanted, and other times it's a solution to a problem they've encountered. Sometimes it's even a shared idea. For me, it wasn't until I met Eric that I found a project that I could really latch on to. This is common, too. With such diversity and possibility in the maker world, it's impossible to predict what other makers might inspire in you, and vice versa (I'll talk more about that in Chapter 8).

Whether it's something you've always wanted to make, or a project or group that sparks your interest at a local makerspace, keep looking until you find the project that lights you up. Coincidentally, it's that internal fire that makes the learning process more engaging and, ultimately, rewarding.

TO THE PROTOTYPERS GO THE SPOILS

With some people, you don't have to dig. When you explain the possibilities and opportunities afforded by the new digital fabrication tools, rapid prototyping techniques and access to makerspaces, they know exactly what they would make. So it was with my friend Eric (a different Eric from OpenROV Eric).

Eric and I were having lunch together just outside of TechShop in San Francisco. I was there working on OpenROV, and his office was a block away. Eric and I hadn't seen each other in months, and I quickly brought him up to date on everything that had transpired with OpenROV—the latest prototype, the Kickstarter project, and how we were building all of the kits. He was intensely curious. With his office next door, he knew all about TechShop but had never gone inside to look around. I explained the tools, with Eric taking a particular interest in 3D printing (as many new makers do). I also made a point to emphasize the ideas of DIT and Just-In-Time learning. It wasn't long into the conversation before I had him convinced that he was totally capable of making anything he could imagine.

I asked him if he had any ideas of something he wanted to make. He didn't say anything, but looked slyly confident as he reached around to his back pocket and pulled out his credit cards wrapped in a thick blue rubber band.

"This." He said. He flipped his tightly wrapped stack of cards around in the air, and explained to me his obsession with finding a wallet that was small enough to fit comfortably in his back pocket. He'd been searching for the perfect solution, trying money clips and thin wallets, but nothing compared to his pleasure of using the elastic band. The only problem was that the rubber band would wear down and break every few weeks. He didn't have any specifics but he knew he wanted to create a better rubber band wallet.

Admittedly, I was a little unimpressed. Sure, everyone loves a thinner wallet but I never imagined a rubber band wallet meant anything more than a piece of blue rubber in Eric's back pocket. I encouraged Eric to try to make it at TechShop, mainly because I wanted to get him in there and using the tools, but I didn't think much about it after that.

It wasn't until a few months later, while combing through different Kickstarter projects, that I would, again, cross paths with the rubber band wallet idea. I randomly stumbled onto a project called "TGT: A New Kind of Wallet"

It was created by Jack Sutter. Jack first got the idea after seeing his roommate using a produce rubber band (taken from a stem of broccoli) as his wallet. He thought it was genius, and he wanted to take it to the next level. Another friend moved into the apartment and wanted to support Jack's idea. She taught him how to sew and helped him make the first prototype of the wallet, which used a thicker elastic fabric with a striped design. Jack wanted to improve the design by adding a little pocket, so he met a furniture designer in NYC who let him take the leather scraps off the floor, which Jack turned into the next prototype design. Later on, after making numerous wallets, another friend helped Jack design packaging for his creations by cutting up an old cereal box. Around the same time, he had created a small logo for his new product, which he was calling TGT, and started sewing them into the wallets.

In a short amount of time, his inspiration had followed the winding and serendipitous path to becoming a quasi-product that Jack and all of his friends loved. The next logical step, of course, was for Jack to put the project up on Kickstarter.

Jack's Kickstarter campaign was a huge success, raising over \$317,000 from more than 7,500 backers. Apparently, there are a lot more people like Eric and Jack in the world (*http://kck.st/19vlLMj*) than I ever could have imagined.

Of course, the first thing I did when I saw the Kickstarter project was send the link to Eric. He was also surprised to hear how successful the project had been. He was a little disappointed, too, that he didn't follow through on his idea, but happy that the idea was finally out into the world.

I now refer to this story as my *elastic wallet moment*. I learned two important lessons from my conversation with Eric and watching the TGT Kickstarter campaign:

Ideas are nothing, prototypes are everything

Eric knew he had a good idea. He even acted on it. He did extensive online searches, scoured retail locations, and bought relevant domain names. He thought about business models and marketing strategies. He went about the project in the traditional business way. Unfortunately, the new maker economy has rewritten the rules.

With the increasing ease of creating (and actually starting to sell) a proto-type, it no longer makes sense to think in terms of the traditional business routine. Of course, once it's up and running, the same accounting, inventory, and manufacturing planning used by more traditional businesses is applicable. But as you travel the M:A spectrum from 1:1 to 1:2 to 1:10, the only thing that matters is a functional prototype.

Prototype first, ask questions later.

Sharing is the new first-mover advantage

Even if Eric had decided to create his own brand of elastic wallets (which he could definitely still do), he'd have a much more difficult time getting the same boost of support as Jack did on Kickstarter. There are only so many thin wallet enthusiasts within the Kickstarter universe, so running a successful project now would be an uphill battle.

It's hard to replicate the ripple effect that Kickstarter creates across the Internet. It's a unique opportunity to magnify your project and broadcast it around the world. Unfortunately, it's an opportunity for *an* idea, not necessarily *your* idea. I've seen numerous projects that failed to pick up traction on crowd-funding websites simply because there was a high-profile, similar project that recently tore through the Internet headlines and sucked all the oxygen out of the blogosphere. For instance, Twine was the first connected device on Kickstarter that aimed to be the platform for the "Internet of Things." The idea was that the rooms and structures around us (and the things in them) should be connected to the Internet, thus they could be measured and controlled. In their Kickstarter video, Twine showed off their prototype sensing the temper-ature of a room, detecting motion and moisture, and other cool features that interested a wide array of backers. They ended their campaign with over $556,541.

After Twine, Kickstarter has seen an influx of "Internet of Things" devices and few have been able to live up to the hype and success of Twine.[1]

The old adage about first-mover advantage was that the spoils belonged to the first product to get to market. In the new maker economy, where the value

---

[1]. Twine certainly benefited from a first-mover advantage, but other Internet of Things companies have still had success. Notably, SmartThings has raised over $1.2 million. Coming in second on Kickstarter doesn't mean you won't be successful, but being first certainly helps.

of your product or company is defined by how others share and contribute to your project and business, the first-mover advantage is given to the first project on Kickstarter.

1,000 TRUE FANS AND 100 TRUE BELIEVERS

One of my favorite documented references to M:A ratios is Kevin Kelly's theory of 1,000 true fans (*http://bit.ly/14ydK4J*). Kelly first articulated True Fandom in response to what he believed to be the artistic aftermath of the long tail, a creative middle class defined herein:

> *A creator, such as an artist, musician, photographer, craftsperson, performer, animator, designer, videomaker, or author—in other words anyone producing works of art—needs to acquire only 1,000 true fans to make a living.*
>
> *A true fan is defined as someone who will purchase anything and everything you produce. They will drive 200 miles to see you sing. They will buy the super deluxe re-issued hi-res box set of your stuff even though they have the low-res version. They have a Google Alert set for your name. They bookmark the eBay page where your out-of-print editions show up. They come to your openings. They have you sign their copies. They buy the t-shirt, and the mug, and the hat. They can't wait till you issue your next work. They are true fans.*

Kelly uses the graph, shown in Figure 7-1, to orient exactly where this middle ground lives in relation to the blockbusters and "the quiet doldrums of minuscule sales."

I love this idea. Not because I think that 1,000 fans is exactly the right number, but because I think it's exactly the right idea. If your game is t-shirts, you're going to have a higher number. If your gig is a highly custom and expensive sculpture, then maybe it's much lower. It's not the number, it's the idea. If your product is revered enough to provide you with a living, it's your job to sift through the other 7 billion people on the planet to find your 1,000 True Fans (or whatever your number is). For makers who aspire to carve out a career with their craft or product, this is a perfect target at which to aim their business aspirations.

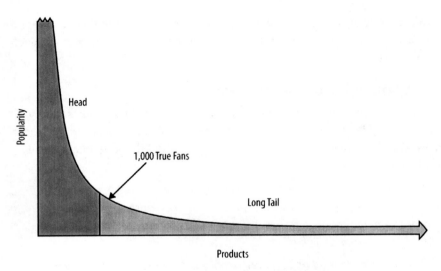

Figure 7-1. 1,000 true fans (based on a graphic by KK.org (http://kk.org/))

As much as I loved Kevin Kelly's blog post, I have one huge problem with true fans as a theory. My problem is that it doesn't look at fandom as a dynamic number, one that changes over time in direct correlation with your skill, experience, and exposure. The True Fan theory, as it stands, is that there are either 1,000 fans who'll buy your work, or else you're stuck. I think the *minimum viable fans* required to earn a living is an important milestone, but I think it's important to examine other milestones along the way. More specifically, the 100 true believers.

Here's my definition:

> A true believer is someone who knows you as the person behind the art or product. Someone you've confided in by showing them your art or explaining your business plan. They care about your product because they also care about you. Not only will they buy your product, but they'll tell everyone they know about what you're doing; they'll get the word out.

The 100 true believers are there before you hit the big time (or medium time). They're the group that knows you personally, sees your budding potential, or that of your project, and wants to contribute to your future success. True fans don't come overnight. It happens one fan at a time until you reach 100 true believers. They are the medium through which you're able to attract and communicate to your true fans. Again, the number isn't important. It could be 100, but it might also be 10. Before you can get to true fans, you have to establish your true believers.

With Kickstarter projects, I think the natural tendency for creators is to spend too much time thinking about the pitch and not enough time thinking about the audience. Not that the pitch isn't important, but more time should be spent thinking about the audience. What is the goal? The goal determines the audience and the audience determines the pitch. To really reap the full benefits of a crowdfunding project, you need to effectively understand where you (as a maker or business) stand in terms of an M:A ratio. Kickstarter can be an effective tool for multiple strategies, but there are two that seem to be most applicable: to develop your true believers or to catalyze them.

DEVELOPING YOUR TRUE BELIEVERS

From the perspective of the casual observer, a Kickstarter campaign for a maker product is very similar to a product launch, an unveiling of a new device or invention to the world. It's the maker version of Steve Job's "And one more thing..." speech. Many naive makers (myself included) have made the mistake of assuming that Kickstarter works that way. They assume that you put up a video and the world beats a path to your doorstep. Unfortunately, that's not quite how it works.

Generally, the surest way to be successful on Kickstarter is to have already developed your true believers. Building your cadre of true believers really takes the hard work and elbow grease out of showing prototypes to the world and soliciting feedback. It doesn't necessarily mean you need to broadcast your project in a public way, like a website or press release, but you do need to do the legwork of explaining your project to friends, friends of friends, at trade shows, and, if possible, to influential people in your respective market. It can be a grind, certainly, but it's the foundation that sets you up for future success.

This period, from an M:A ratio of 1:1 to 1:100, is critical. It cuts both ways, too. It's an important time to find the first enthusiastic supporters of your idea and product, but it's an equally important time for you as a maker to grow into understanding exactly what the market will support. It's co-evolution; the world learns about you and your product, and you learn about yourself and the best way for your product to fit into the world.

Behind many of the seemingly overnight successes on Kickstarter are stories of a maker (or maker team) who worked incredibly hard to perfect her product.

Before Abe and Lisa launched their wildly successful Kickstarter project for Nomiku, they spent over a year on the maker circuit: showing off their DIY sous vide device at Maker Faires, offering kits and classes on how to build the kit, and continuing to improve and evolve the product. By the time they launched, no one on the planet knew more about low-cost sous vide machines and techniques. They knew Nomiku was truly innovative and they had built a community of friends and supporters who immediately got behind the project.

When I first met Anton and saw his Oru Kayak almost six months before he launched his project on Kickstarter, I begged him to let me take it out for a test ride on the San Francisco Bay. We spent an afternoon paddling around Berkeley and I put the kayak through all the paces I could imagine. For a folded kayak, made from the same material as the political ads you see staked out in front yards, I was really impressed! I told Anton I wanted to write about my Oru Kayak experience on the *Make*: blog. He was hesitant, as he wanted to save any press opportunities for his upcoming Kickstarter project. I finally convinced him to let me write the story, and it ended up driving a lot of traffic and attention to his site and project, and he collected email addresses of those who expressed interest. By the time he launched his Kickstarter campaign, almost five months later, he told me that getting the word out early was one of the best things that could have happened.

To be fair, every once in a while, a Kickstarter project without the 100 true believers will break through to be a smashing Kickstarter success, like the PrintrBot. More likely, though, the success of the project is closely tied to the hard work and networking of the team prior to launching the project.

CATALYZING YOUR TRUE BELIEVERS

Kickstarter or Indiegogo or any other crowdfunding platform isn't the only way to turn your project into a business. But for those artists and projects that are ready, it's a great way to go from true believers to true fans in short order. In my mind, this is the true genius of the Kickstarter model. It's much more fun to watch your project spin around the Internet by word of mouth than spend a month repeatedly having to remind your friends that you need their support.

As a thought exercise and gut check for knowing where you line up on the M:A ratio, try to make a list or map of 100 people that know what you're doing and have expressed enthusiastic support (again, the right number might be 100 or it might be 25; it depends on the project). You might even send out a few emails telling those people that you're preparing to launch a Kickstarter project and you'd love their feedback or input.

For our OpenROV project, the Kickstarter campaign was a long time coming. Eric and I started the project over a year and a half before our campaign. We invited anyone and everyone to join our community of DIY underwater explorers and enthusiastically shared our plans and designs. By the time we launched our Kickstarter campaign, we had collected a list of over a thousand people who had signed up on our site and expressed interest in building their own OpenROV. For us, Kickstarter was about getting to the next level by catalyzing our true believers. By taking the time to build support and involvement before the Kickstarter project, we were able to reach our funding goal within a few hours, which came almost entirely from the first email we sent to the OpenROV community.

Seth Godin, who ran a successful Kickstarter campaign for his book, came to a similar conclusion:

> Kickstarter appears to be a great way to find fans for your work. You put up a great video clip and a story and wait for people who will love it to find you.
>
> But that's not what happens. What happens is that people who already have a tribe, like say the "punk cabaret" musician Amanda Palmer, use Kickstarter to organize and activate that tribe. Kickstarter is the last step, not the first one.

He's almost right. Done correctly, Kickstarter isn't the first step, but it's also certainly not the last. It's in the middle—the start of something new. It's the beginning of having true fans, which comes with an immense responsibility to serve and deliver. It's a lot of work. It's the best imaginable type of work: co-creating with a group of people who share a common vision.

For OpenROV, the best part of the process has been exactly that. We're working exclusively with them to manufacture and distribute kits to anyone interested in contributing. We found our True Fans only because of the amazing support from our True Believers.

Regardless of where you stand in terms of fandom, the first step of a successful crowdfunding campaign is honestly assessing your M:A ratio.

PROTOTYPES AS PRODUCTS

The maker movement is doing more than organizing a community of manually-literate collaborators and reducing the costs and barriers to accessing quality digital fabrication tools. It's also creating a new swath of educated *prosumers* (producers + consumers) who place a higher value on the transparency and hackability of a product as much as they do on the price and performance. They care about the product *and* the process.

This new maker market is also more forgiving if a product (or project) isn't polished or packaged beautifully. Instead, they're more concerned with participating in the ongoing development. In addition to providing feedback and ideas for improvement, this maker market is also eager to buy up the in-process goods. Essentially, it's a market for evolving prototypes.

The old model of manufacturing and product development involved an innovative idea, expensive prototyping with designers and manufacturers, marketing budgets for the launch of the product, and expensive distribution channels to deliver the product to market. The maker movement has erased those barriers. More aptly, it's rolled all of those steps into one public process that doesn't take the intense capital expenditures inherent in the old manufacturing model.

The innovative idea is still a necessary catalyst, but the path in the new maker economy quickly diverges from there. Whereas the prototypes in the old model serve as milestones and expenses toward a finished and polished product, the maker model turns the prototyping process into a product and a social object used to educate and build community. The maker's challenge is not to build a finished product, but to create something that is good enough and capable of getting better; the maker is designing for a perpetual state of becoming.

For more complex devices or inventions, this proto-product often takes the form of a kit. Chris Anderson and his 3D Robotics company got their start selling the ArduPilot boards and moved into quadcopter kits. For MakerBot, the first desktop 3D printers that they shipped were all kits that users had to assemble themselves.

The proto-product method has many advantages:

It's less expensive

On the business backend, creating and packaging a kit is far less labor intensive (and therefore less cost intensive) than creating a finished product. It's possible to run a kit business with just a few people, as opposed to needing a manufacturing partner or assembly line. The kit business will have higher customer service demands, though, so it's important to build the platform so the community can help educate one another. A wiki with thorough build instructions and an accessible FAQ section goes a long way in answering many of the most persistent questions.

You learn more

Undoubtedly, people who build your product themselves will come up with different strategies, whether that's due to education, available resources, or just sheer happenstance. The diversity of creation creates a Darwinian process that susses out the most effective way to build and use the product. To rehash the Michael Schrage quote from Chapter 3:

> Talented amateurs don't just build kits; kits help build talented amateurs. And healthy innovation cultures—and successful innovation economies—need the human capital that their talent embodies. Kits are integral, indispensable, and invaluable ingredients for new value creation.

Low barrier to entry

It doesn't take huge capital expenditures to start a kit business. Especially with the presale method of Kickstarter, you can get your business off the ground with very little cash at the outset. It's easier to budget the costs and expenses of running a business when the Bill of Materials (BOM) remains exactly that: raw materials. Adding in the real estate, insurance, and kit-packing labor costs will give you a "good enough" idea of the economics that make your business work.

There's also a growing outlet and distribution channel for well-designed and popular kits. In addition to marketing through Kickstarter, companies like Adafruit and Sparkfun as well as *Make*:'s Maker Shed are becoming effective distribution channels for makers to get their projects out to a wider audience.

Going from zero to one and then from one to one thousand has never been more possible for makers like you and me. But getting to 1000 true fans is still a monumental task. Even though the process is made easier with the new digital fabrication tools and Internet distribution models, it's an immense challenge (not to mention the new challenges that arise when product demand increases and orders start to number into the thousands).

Makers are becoming entrepreneurs, and in true maker style, they are going about it in a radically collaborative way.

8

Makers Going Pro

Having a sudden surge in demand for your product is both wonderful and stressful. It's great to have the new business, but there are usually growing pains associated with the added pressure and attention. For maker businesses—especially those who are on their own or part of small teams—this sudden surge can be jarring.

I observed firsthand how one team of makers, 8bitlit (*http://8-bit-lit.myshopify.com/*), dealt with this overwhelming surge. The experience provided insight into how these maker groups are handling the jump from product to business. It also gave me a preview of how this whole maker economy might look for entrepreneurs as well as employees.

In March of 2012, the *Make:* blog ran a feature on Adam Ellsworth and Brian Duxbury's delightful Coin Cube, a yellow box made of laser-cut acrylic and equipped with a sensor array and LED that lit up when you "punched" the bottom. The creation was a real-world homage to cubes seen in the Super Mario Brothers video games. The focus of the *Make:* write-up was exploring how entrepreneurs and makers meet at makerspaces and go on to create businesses. One of the side effects of the article was a huge jump in attention. The story of 8bitlit and their delightful cube lamps rippled through the Internet, garnering attention on many popular blogs and gaming websites. The orders for their kit increased along with the traffic, and Brian and Adam quickly had more work than they could handle on their own. The problem was compounded by the fact that Brian had recently started a new full-time job, leaving an even larger burden on Adam's shoulders.

I was working out of TechShop in San Francisco at the time, where a Coin Cube hung over the entrance to the work area as I toiled away on my re-skilling quest and the latest OpenROV prototype. I was fortunate enough to know Adam quite well during that time. I was delighted to see the well-deserved recognition and fascinated to watch the effects that all the publicity (and hundreds of new orders) had on their fledgling business.

I watched Adam and the team from across the room as they slowly (but respectfully) took over almost the entire shop with laser cutting, silk screening,

CNCing, and packaging the cubes. There was a visible expansion that took place; we could all see that things were going very well. But what was more interesting to me was the rallying of the TechShop community around Adam's success, with numerous other members chipping in to help with various steps in the process. During the busiest month, Adam estimated that as many as nine different Tech-Shop members and staff had contributed a total of between 150 and 180 man (and woman) hours over the past few weeks. Heck, I even chipped in for an hour of screenprinting.

"We never would have been able to get this done without the community here," Adam told me. "We got three hundred orders over a three-week period, and only a two-week lead time for delivery. It would have been impossible."

Of course, this type of situation happens all the time—businesses hire tempo-rary workers to fill surging demand. But this story had a maker twist.

"Everyone is capable of doing any task—silk screening, electronics, laser cut-ting, whatever. I think that's specific to the fact that we're getting people from TechShop," Adam said. "It's not that everyone knows everything, but they're all comfortable learning the different machines. They just learn from each other."

It wasn't just a good situation for Adam, either, or just some fluke of goodwill. I talked to some of the TechShop Members who had been on the Coin Cube pro-duction line. Sam Brown had been designing and working on a board game, Lyssan, that was recently funded on Kickstarter. While his boards were off to the printer, Sam found himself with some extra time. After seeing a flyer that Adam posted offering hourly work and seeing the growing work area that the Coin Cube was taking up, he decided to join the fun. During that month, he worked over 30 hours for Adam.

"It's great to be working with other entrepreneurs. Even though our products are in different categories—electronics and board games—I'm still learning a lot from watching him go through this process," Sam explained. "There are similar issues that I will face, like shipping, keeping customers happy, stuff like that."

Alex Glowaski joined TechShop less than a week prior to the cube lamp bo-nanza. Alex was unemployed at the time and had joined TechShop because she wanted to find more hands-on work and also work on her own projects. She had recently finished a cool, wearable transit card and told me about new ideas for some wearable holograms she wants to experiment with. Because her projects didn't have a business model behind them yet, she was happy to pick up work with Adam's team, putting in about 25 hours during that period.

"Pretty much everyone here has a Kickstarter for something," Alex told me. I think she was right. Almost all the makers I've met, at TechShop or elsewhere, are in some stage of the Kickstarter process: planning, campaigning, or fulfilling. Even though the Coin Cube didn't go the crowdfunding route, the overwhelming demand spikes and the sudden workload that ensues are very similar to a maker Kickstarter project.

The Coin Cube story encapsulates the best features of this world. In the new maker economy, a makerspace membership is the new entry-level job. We're teaching each other how to make it all work. The companies are collaborative and productive.

The makers are going pro.

Allowing for a Shared Vision

The story of 8bitlit (Adam Ellsworth and Brian Duxbury) began as one of friendship and creativity. Travis Good, the writer who penned the original story on *Make:*, had an agenda. Travis had spent the previous year traveling around the country to dozens of maker- and hackerspaces, even co-founding one in Washington, DC. His curiosity and passion for the maker movement eventually led to a gig writing for the *Make:* blog. His focus: how people meet in makerspaces, how projects and collaborations evolve, and how, sometimes, maker businesses eventually emerge.

Having visited so many different makerspaces, Travis was able to gain a unique perspective on some of the larger trends in this collaborative new world. He decided that these co-creation stories—makers serendipitously meeting and creating projects together—was the most important issue he could report on. Adam and Brian were the first team he highlighted, but it didn't take long to fill up a list of examples: Phil Torrone and Limor Fried collaborating to create Adafruit; Zach Smith, Bre Pettis, and Adam Mayer creating MakerBot at NYC Resistor; Eric and I meeting through a BioCurious connection; Abe and Lisa Fetterman and their DIY sous vide machine.

The list goes on and on. Of course, there are equally as many examples as makers building something that they always wanted themselves, like Anton Willis and the Oru Kayak, but many of those lone founders would tell you they wished they had had a teammate.

If you're just getting started, stay open to the idea of joining a team and collaborating to create something bigger than any one maker could accomplish alone. If nothing else, it's a more fun and engaging way to move along the maker path. I can say without hesitation that working with Eric and the rest of the OpenROV community has been the most rewarding experience of my life.

First Who, Then What

The story of 8bitlit didn't end at the surging demand of Coin Cubes. In fact, that was just the beginning. During that stressful order-filling month, the team—Adam, Sam, Alex, and Ryan—realized how much they enjoyed working with one another. They discovered that they each brought a different set of skills to the table and that together they comprised a well-rounded maker dream team.

They decided that they would continue on as a unit, although unsure of exactly what they would create next. While continuing to fulfill orders for the Coin Cube, they set their sights further. They didn't want to be tied down to any specific product, instead becoming something of a maker factory for new ideas. Thus, the new venture ProtoTank (*http://www.prototank.com/*) was born.

The quartet of makers, plus or minus a few part-time helper members for odd jobs, took up a small office in the back of TechShop. They hung a small sign on the door and were open for business. In addition to the Coin Cube order fulfillment, they started tinkering with new projects. They hacked into a Sphero, a small, iPhone-controlled ball, and tried to get it to swim. They attached a smartphone to a six-legged robot toy to create a cheap, remote-operated hexabot. In just a few short months, their office was filled with gadgets and parts from nearly a dozen side projects the group had been working on.

As an outside observer, I couldn't help but marvel at the group's creativity and sheer output of doodads and gadgets. I could always walk into their office if I needed a spell of inspiration or if I was missing a resistor or LED. Most important, though, they were having a lot of fun, enjoying each other's company and feeding off the creativity of the group.

About six months into their maker experiment, they came up with the product that became their sole focus: color-changing acrylic signs. They had been experimenting with a new material, color-changing acrylic, when they discovered that they could create really unique signs that used perimeter LEDs to light up a design or word in a clear sheet of acrylic plastic, giving the appearance that the lighted words were floating in the middle of a window. The words were visible when the LEDs were turned on but transparent when turned off. It's quite an intriguing sight.

The group decided to pursue a patent for their creation and are now moving full-speed into product development and production.

But even if the sign idea doesn't pan out, the ProtoTank team isn't worried. They have a long list of product ideas to try next. They are the epitome of Jim Collins' famous advice in his book *Good to Great* (HarperBusiness, 2001): first who, then what. Collins explains that the best teams are focused on finding the right people first and then deciding what to build and create. Finding the right maker team or partner is no different—it's all about the people. On the day that Eric and I first met, and he told me the story of the Hall City Cave, we spent almost the entire afternoon discovering our shared values of adventure and exploration. The idea of OpenROV actually becoming a business didn't come until much later, and only then because we saw it as the best way to share our enthusiasm with more people.

Paying the Bills with Side Jobs

Despite their wide-open development strategy, ProtoTank was very pragmatic in the design of their operation. They gave themselves a *runway*, a startup term that translates into the amount of cash a company has in the bank and how much corresponding time they have to gain traction with their product in the marketplace. The typical strategy for building a runway is to raise money from angel investors or venture capitalists. ProtoTank didn't do that.

They built a runway by taking on side projects and contract work for other groups that needed stuff built. Rapid prototyping and making skills are in high demand, and ProtoTank realized they could build a steady stream of income— enough to keep them going indefinitely—by taking on a percentage of the many offers and contracts that seemed to find them.

It wasn't part of their original plan, either. They thought they would need some kind of venture funding to get the company off the ground, but were quickly overwhelmed with offers and projects from others who wanted to utilize the diverse talent of the ProtoTank team. They found themselves building large installations —race tracks and robots—for events and conferences. They found work creating LED jackets for companies that didn't have enough makers on staff. They were even tasked by major corporations to try to "hack" products, taking apart devices in the hopes they could find a new use or purpose for the product.

Although the side jobs were never in the business plan, they have given ProtoTank the freedom to experiment. They're diligent about keeping a large percentage of their time for their own products and projects, but can also employ patience in their product development process because it isn't their only source of revenue. They keep overhead low and costs down by renting space inside of TechShop, and they also use the community as a platform for meeting new clients. ProtoTank is the clown fish of the TechShop coral reef: they've become a critical part of the maker ecosystem.

Side jobs and part-time work seem to be the main ingredient in keeping the TechShop ecosystem (and wider maker economy) healthy and dynamic. Normally, such reliance on part-time work would be cause for anxiety, but at TechShop it seems to keep the situation fluid and receptive to demand shocks.

The lessons of ProtoTank are applicable for any new maker, too. As opportunistic as it might seem for the companies and micro-manufacturers, the side gig economy can also be an incredibly freeing opportunity for the temporary worker. With everyone moving around and working where there is demand, the sense of entrepreneurialism is always present. Everyone has their own project to work on. If your project becomes big enough to need full-time attention, it's easy and natural to transition into giving it the time it deserves. That's when you walk the other side of the maker economy line, giving people work as they learn the skills and get their own projects off the ground. It's a beneficial cycle of making and helping each other build products and companies. It's Do-It-Together company-building.

Maker Business School

What started out as friendly advice and support from other maker businesses is quickly turning into a more formal infrastructure. Accelerator programs are springing up that house early-stage startups, providing funding (usually $20,000 to $50,000), mentorship, and access to workshop space to help these makers navigate the uneasy transition from hobby to company.

The accelerator model for startup incubation has been around for many years. Popularized by Paul Graham's Y Combinator, the hypothesis was that the barriers to entry for software startups are so low that it makes sense to gamble small amounts of money on a talented team with a set timeline to try to gain traction for their product idea. Many popular and valuable software companies, such as AirBnB and Dropbox, have since emerged from these programs. And the success of Y Combinator has created a boom in similarly modeled accelerator programs around the world.

Hardware startups—companies that make real, physical things—were traditionally left out. Investors were scared off by the large capital investment and the myriad things that can go wrong in the manufacturing process. Hardware was just too hard for many investor appetites. But the maker movement has changed that. Increasingly accessible and affordable prototyping equipment coupled with a maturing understanding of the manufacturing process (especially in Shenzhen) and the popularity of crowdfunding sites like Kickstarter has created a similarly low barrier to entry like the one software startups enjoy.

In late 2012, Paul Graham, the founder of Y Combinator, described the changing investor perspective (*http://www.paulgraham.com/hw.html*):

> Investors have a deep-seated bias against hardware. But investors' opinions are a trailing indicator. The best founders are better at seeing the future than the best investors, because the best founders are making it.
>
> There is no one single force driving this trend. Hardware does well on crowdfunding sites. The spread of tablets makes it possible to build new things controlled by and even incorporating them. Electric motors have improved. Wireless connectivity of various types can now be taken for granted. It's getting more straightforward to get things manufactured. Arduinos, 3D printing, laser cutters, and more accessible CNC milling are making hardware easier to prototype. Retailers are less of a bottleneck as customers increasingly buy online.

And right on cue, we're seeing an influx of new hardware startup accelerators. But hardware scales differently. For a software company, going from a thousand users to a million is a matter of server space and code. For a hardware company, it's an entirely different process and supply chain. It's a dramatically different company. As such, the hardware-focused accelerators are providing different services than their software-based counterparts. Lemnos Labs in San Francisco offers up to $50,000 in funding to their startups, much higher than the typical accelerator seed amount. Their thinking is that prototyping hardware is more expensive than just feeding and housing the engineers. Real prototypes take real materials, which takes real money. Haxlr8r has based its program in Shenzhen, betting on the fact that being close to the beating heart of global manufacturing will give its entrepreneurs a leg up on the competition and a head start on the manufacturing process. Bolt, a Boston-based design accelerator, is helping its entrepreneurs license their products in addition to just building companies around them—a different take on business opportunity.

Even major league manufacturing companies are betting big on maker startups. In 2013, PCH International, one of the global leaders in supply chain and product development consulting, opened a new 30,000 sf facility in San Francisco (*http://techcrunch.com/2013/05/24/mr-china-goes-to-san-francisco/*). With most of its operations based in Shenzhen, the new space in California was created to give entrepreneurs in the United States the same consulting and design services that PCH normally offers to clients in Shenzhen. The space boasts a 7,000 sf state-of-the-art rapid prototyping lab as well as a full staff of development and design professionals. It serves its Fortune 500 clients, but also serves as an accelerator to help maker businesses scale up production.

As much as I appreciate the human scale of the new artisans, sometimes this type of scaling is what these maker businesses need to do. When your Kickstarter project soars above 3,000 units or you have a demand from a large retail partner, there's suddenly a lot more on the line and a very real pressure to produce, and produce quickly. These accelerators and incubators are a great bridge to go from maker to manufacturer. Of course, there are going to be growing pains and challenges, but at least there's a group of mentors and advisors that can help you cross the most turbulent waters.

When I talked to Lisa and Abe about their experience with Haxlr8r, they couldn't say enough good things about it. It was difficult and stressful, of course. Starting a company always is. However, this was exactly the kind of pressured environment they needed in order to move from a DIY kit-selling side project to a full-scale (and full-time) business. It gave them the opportunity to make the leap.

And it might be a good opportunity for you to make the leap, too. Again, from Paul Graham's essay on hardware:

> *So if you want to work on hardware, don't be deterred from doing it because you worry investors will discriminate against you. And in particular, don't be deterred from applying to Y Combinator with a hardware idea, because we're especially interested in hardware startups.*
>
> *We know there's room for the next Steve Jobs. But there's almost certainly also room for the first <Your Name Here>.*

APPLYING TO A HARDWARE ACCELERATOR

What's it actually like applying to one of these programs? Is it like applying to college or like applying for a driver's license? What skills are they looking for? How refined should the idea be?

I asked Cyril Ebersweiler, the co-founder of Haxlr8r, what he looks for in an application. After seeing their demo day in San Francisco, I found it difficult to piece together any trends in the different companies, other than the fact that they were building hardware. After our discussion, I realized I wasn't very far off. It's a moving target for Haxlr8r as well. In only its second year, the company is still figuring out what works and what doesn't. Instead of "choosing" projects, they see themselves as partners. They look at each situation and project how much value Haxlr8r would be able to provide, in terms of help with product design, preparation for Kickstarter, or setting up supply chain partners.

Even in the short amount of time they've been experimenting, Haxlr8r has learned some important lessons. Those lessons are an important lens in how they evaluate the different applications. Here were some pieces of advice that Cyril gave me:

Have a team

Haxlr8r has had two companies that consisted of just a sole founder. And although Cyril said they would consider possibly doing it again, it would have to be an extreme case. It's just too difficult to go through this process alone. When you're growing and, more important, iterating quickly, there is so much parallel processing that needs to occur. It's very difficult for one person to manage this by themselves. Eric and I can attest to this, too. There's just no way we could have gotten OpenROV off the ground if we didn't have each other (not to mention the support of the entire community).

Know your skills

You can't just put together any team, either. Cyril explained that Haxlr8r also looks for diversity among teammates. They want skills that overlap: software developers, mechanical engineers, business-minded people. It's the blending of the unique abilities that sets great teams apart. The self-knowledge to recognize what each individual is good at combined with the ability to divide the work into those domain specialties is a hallmark of a high-functioning team. Nonetheless, it's also important to be flexible. Cyril likes to see overlap on different responsibilities, so different team members can pick up the slack whenever work piles up in one area. So even if your specialty is mechanical engineering, you also need to have a handle on the business side of the company, and vice versa as much as possible.

The intersection of software and hardware

In terms of product ideas and industries, Haxlr8r is smitten with the interaction between hardware and software. The *Internet of Things* has become a popular term among investors referring to devices, household and otherwise, that are being built to connect to the Internet. It is creating an entirely new class of devices. The Nest Learning Thermostat is one of the best examples: it takes a device, in this case the household thermostat, and completely reimagines its capabilities. It remembers your habits and preferences, can be controlled from anywhere through your smartphone, and automatically adjusts when you're away. The technology creates a more pleasant environment as well as significant savings on energy bills. And that's just one example of the Internet of Things.

Brad Feld, a partner at the Foundry Group, commented that his fund is very interested in "software wrapped in plastic" (*http://bit.ly/18ooqdy*) when explaining his investments in maker companies like MakerBot. The same intersection of software and hardware that is enabling so many new makers is also catching the eyes of investors.

Robots

Cyril also said that Haxlr8r is interested in anything and everything robotic. In the same vein as the software and hardware intersection, there are powerful forces at work driving a new era of robotics. Driven by the demand for smartphones and tablets, sensors and components are becoming commodity items, easily reconfigurable into new and exciting uses. The booming personal drone market is a perfect example of an emergent and derivative industry, stemming from the price pressures on smartphones and tablets.

Resume Builder

Freelancers, free agents, creative class, independent workers—whatever you want to call them, you know the type. Maybe you are the type. Working in and out of coffee shops, chasing down clients, jumping from project to project. It seems like more and more people I know are making their living as some type of a contract worker. Some of them do this by choice, like web or graphic designers, because they like the flexibility. Others have been thrown into the fray due to the turbulent economic climate.

For the past decade, I've been in awe of these people. I always envied their freedom and resilience, and confidence in their creative abilities. I always wanted to be able to put web design, graphic design, Photoshop, and other creative skills on my resume. I still do, but I've never been confident in my ability to create anything myself, let alone ask someone else to pay for it. The freelance economy is too competitive for me to try to get this type of work. I am certain my skills are far enough behind the average freelance designer's that it would take me years to catch up.

But imagine learning web design in the 90s. The web was so young, so new. The opportunity was immense, but it was also fairly straightforward to learn all the skills you needed to play in this new sandbox. The tools were trivial compared to the complex and competitive Internet economy we have today, and you could basically learn on the job because *everyone* was getting up to speed. Today, many of those same skills are seen as prerequisites. The people who took advantage of that moment in time—those who learned the digital skills that underpin the Internet economy—were able to create tremendous opportunity for themselves. Whether they knew it or not, they had opened up the doors to all sorts of interesting career options. Whereas those of us who didn't have been racing to catch up.

Now I finally have one skill that puts me on the cutting edge of freelancing: rapid prototyping.

I once heard my friend, Andy Lee, a design engineer, describe rapid prototyping as a fancy term for quickly trying a bunch of ways to do something, most of which won't work. The goal is to use as little material and time as possible to try to hit on an idea that works or solves a problem. Then, when you find a promising solution, you continue to prototype and evolve a design.

The goal of rapid prototyping is to show a physical proof-of-concept. The skills and knowledge you develop by following the steps laid out in this book have already put you in the top one percent of rapid prototypers in the world. Harnessing powerful tools (3D printers, laser cutters, basic CAD design software) and knowing how to navigate the DIT maker world (finding designs on Thingiverse and collaborators at local makerspaces, utilizing online forums) can amplify your productivity to a point that would make any employer or client impressed.

The next few years are an incredibly opportune time to pick up these skills and add them to your resume. It's akin to the early 90s web developers and designers. It was very difficult to see just how far-reaching and door-opening that suite of web skills would be at the time. It's similarly difficult to predict where this maker movement will go and what opportunities it might inspire.

Here are some ways to give your resume a MAKEover:

Added skills

The easiest way to show off your new skillset is simply by listing them, right next to speaking Spanish or video editing or kiteboarding or whatever else you put in that section of a resume. I'd list it like this:

Rapid Prototyping (Physical Products)—Working knowledge of digital fabrication tools like 3D Printing, Laser Cutting, CNC Machining, 3D Scanning.

Of course, only list something that's true. Adding this dimension—the skills to make actual, physical prototypes—will certainly make your resume stand out from the rest of the digital natives.

List your projects

The big problem with resumes is that they only show what you've already done, not what you'd like to do next or what skills you'd like to continue developing. Whether you're just graduating from college or trying to make the switch from an entirely different career, it can be difficult to know how to effectively convey what you'd *like* to be doing next. Having the right experience can be a "chicken or egg" problem—you can't get more of it, because you don't have enough of it.

Fortunately, if you want to transition into a maker job, there's a way around this outdated formality: start making things! List the projects that you've worked on and completed instead of talking about previous "responsibilities" you had at a job. In the new maker economy, it's much more impressive to list what you've created rather than a title you've held.

No one is stopping you from starting a new project. And it doesn't have to be a big, Unknown Project like we talked about in Chapter 3. Even the kits and Known Projects you've worked on are an impressive thing to list. If you've put together an OpenROV, built an ArduCopter, or built your own RepRap 3D printer, put that down!

Under-development projects count, too!

I noticed one characteristic of makers very quickly: they have a dozen side projects going at any given time. Projects that are on the back burner. Projects they recently started. Projects they were working on months ago and have been meaning to get back to. Even the completed projects usually have a final tweak or follow-up coming. Nothing is ever finished; everything is in a constant state of "becoming."

If you're working on something and have made any sort of measurable amount of progress, don't be afraid to talk about it or list it. As someone who's looked at resumes for makers, I can tell you that I look for a breadth of different projects.

No Experience, No Problem

Whatever you do, don't pretend to have skills or knowledge that you don't actually have. Don't even slightly exaggerate it. You're only going to get yourself into trouble.

You'll almost certainly find yourself in an uncomfortable situation where you'll be expected to know something you don't.

But more important, you're missing out on what is potentially your biggest asset: inexperience.

Instead of thinking about inexperience as a disqualifier, try framing it in terms of a competitive advantage. If you're clear and transparent about your lack of knowledge and can articulate a desire and a plan to try to learn, that can be a powerful asset to a maker business. Every maker business is trying to reach more people. Oftentimes, those new customers are going to be just as inexperienced as you are, so seeing the product or community from the perspective of a complete newbie can be wildly useful.

This is how I got my start with the Zero to Maker column. Instead of hiding my insecurity about my manual illiteracy (which terrified me) and staying on the sidelines, I turned that into my story. I freely discussed my desire to learn from scratch: What should I learn? Who should I talk to? What don't I know?

The process was illuminating for me, and it gave me a great foundation on which to build new maker skills. Coincidentally, the process of blogging and writing about my journey created a resource and map for other new makers to follow.

The same strategy could be used for anything: 3D printing, building and programming drones, welding. Whatever you want to learn, there is probably room for a resident newbie in the space. It might not always be a job opportunity (or even directly lead to a position), but it will certainly give you a platform on which to learn and give you a better chance at getting a job in that domain. I'm willing to bet that this type of DIY education will end up costing a lot less than any traditional program you could attend.

Maker Businesses Need More Than Just Maker Skills

As products or projects make the transition into actual businesses—increasing sales and demand, successful Kickstarter projects, or even venture capital investments—makers are having to "go pro." Running a business efficiently and effectively can be much different than making a product or prototype. In many ways, a maker business is a lot like a traditional business, whether retail, corporate, or startup. Skills and responsibilities such as customer service, managing shipping and fulfillment, accounting, blogging, and social media are all roles within maker businesses, too.

As I've discovered personally with the OpenROV project, these more traditional business or job skills take up a lot of the actual work you must do every day. For many makers-turned-entrepreneurs, this is the type of work that we're not particularly good at (and therefore need the most help with). Not coincidentally, this need creates an opening for someone who is looking to build a career in the new maker economy to get a foot in the door.

To give you an idea of what I mean, here's a snapshot of the jobs being advertised by MakerBot Industries in Brooklyn, New York, taken at the time of this writing (December 2012):

Current openings in Brooklyn, New York:

ADMINISTRATIVE
- Office Manager

FINANCE
- Senior Procurement Manager

MARKETING
- Traffic Manager
- Senior Designer
- Senior Product Marketing Manager

PRODUCT
- Software Engineer: Computational Geometry/Image Processing
- Software Engineer
- Quality Assurance and Test Engineer
- Electrical Engineer

RETAIL STORE
- Retail Operator for MakerBot Store

SALES
- Inside Sales Representative

OPERATIONS
- Director of Operations

SUPPORT
- Support Representative

HUMAN RESOURCES
- Human Resources Administrator

BUSINESS DEVELOPMENT
- BizDev

As you can see, they're growing quickly, with sixteen different positions listed on their careers page. There are a wide variety of roles within the company, ranging from director of operations to support representative. Of the sixteen jobs listed, only four of them are engineering positions. Other jobs, such as office manager or inside sales representative, don't require an extensive engineering background. These roles are critically important to the continued growth and expansion of MakerBot, and they are a great opportunity for new makers to jump into the action. At the same time, makers can put their skills and experience to productive use. If you take a sales job at MakerBot, you're putting yourself in the center of the new maker economy and surrounding yourself with makers from all over the world.

Your sales experience or office management experience can be your ticket into a new maker career.

Here are some of the most useful skills to apply to a job in the new maker economy:

Writing

If you look behind the curtain at the fastest growing companies in the maker movement, you'll likely find a writer or media person behind it. Chris Anderson, founder of DIY Drones and CEO of 3D Robotics, is a writer at heart, having spent over eleven years as the editor in chief at WIRED Magazine and writing three books during that period. The skills he honed as an editor, generating and filtering content from a community of contributors, turned out to be directly applicable to open-source hardware development. Instead of blog posts and magazine articles, the outcome of this curation has been the technology suite of flying drones and quadcopters.

DIY Drones isn't unique. Phil Torrone, Creative Director at Adafruit, is a writer who has written for *Make:* and also started the website Hack a Day (*http://www.hackaday.com*). Bre Pettis, CEO of MakerBot, is very fluent in new media and has spent a lot of time generating how-to project videos for the Internet.

We experienced this first-hand with OpenROV when I started blogging about my Zero to Maker experience. Writing about my quest—the process and the tribulations—was the genesis of much of the attention and growth. It helped us gain attention, but more important, it helped us to refine and effectively communicate our message to the world. We learned how to tell our story.

If you have a bent toward writing and communicating, there is space for you in this maker world. Making is 50 percent building and 50 percent sharing and communicating. Every maker project, from pre-Kickstarter prototype to fast-growing company, has room for someone who can help them communicate more effectively.

Customer service

The number one surprise (accidental) entrepreneurs discover during success-ful Kickstarter campaigns is overwhelming customer service demand. Britta Riley, the creator of the popular Windowfarms project, commented on one of my *Make:* blog posts (*http://bit.ly/14Y5f9i*):

> When you have customers numbering in the thousands, it is most assuredly a different story from packing boxes and an-swering emails in your garage. You start to need things like en-terprise level software for data management because the crowdfunding sites do not provide basic ecommerce support. We are a company that is committed to transparency and lives by the motto, "release early release often," but at some point we were faced with the realities of the $$ cost of transparency and frequent publishing. Updates on Kickstarter take someone's time and energy to craft and you really want to be certain about what you're promising. We found that every time we made a post, it created huge customer service loads. With each post, we would get a flood of hundreds of emails from people who wanted to change their address, ask us a question about their specific cases, or even just rant about how awful we are as a company for being late and let us know they were reporting us to the Busi-ness Bureau. Customer service became a full time position that we had no way of estimating in the advance.

It's all the little things that you need to consider: How much will shipping cost to this or that location? Can your product do this? Would I have any trouble if...?

If a project attracts enough attention, you can be certain that a steady stream of comments and questions is going to follow. There are a number of ways to help cope with the flood, but there will always be a segment of people who don't read the FAQs or don't have the patience to post the question to the community forums.

Many maker entrepreneurs and fast-growing maker businesses are strug-gling to keep up with the customer service demands that their product is gen-erating. Even for projects that rely on an active and engaged open-source com-munity like DIY Drones or OpenROV, the questions about shipping, kit availability, and some technical issues are always going to have to be solved by someone on the business side.

The same customer service skills learned at a previous job can be directly applied to a job with a maker business, and will give you a financially sustainable opportunity to spend more time around makers.

Operations—shipping/sourcing/inventory

As much as a maker business might start to look like a traditional web startup, running it is still a function of operations. Supply chain management is the backbone. With small-batch manufacturing, any excesses in inventory or supply costs can quickly run an operation off the tracks. No matter what quantity you're making, as soon as you get over 100 units, all sorts of challenges and unexpected setbacks start to kick in.

With OpenROV, I knew it was going to be a challenge, but I never anticipated it would be the *entire* challenge. Our margins were small enough that any significant mistake could have seriously derailed us. Luckily, we had a superhero come to the rescue. Zack Johnson, the same TechShop Dream Coach who helped guide me through the first few classes and skill-builders, turned out to be an ideal ally in our battle for supply chain victory. At the time, Zack was running the retail store at TechShop and was managing all the relationships for parts, supplies, and materials for the shop and all the classes. He had built up a network of the best resources in the Bay Area, and he could navigate a Grainger Catalog or Digikey website like nobody's business. Zack knew about local acrylic suppliers we had never heard of, and he was on a first-name basis with them. He helped us get better materials, with shorter lead times, and at a lower price. In all, he probably saved us about $15,000 on our Bill of Materials. This is not insignificant when you consider that our entire Kickstarter raised $110,000.

Managing inventory, running a shipping department, or working on an assembly line are all skills that could help a growing maker business.

Licensing

Inventing or creating a product is very different from starting and running a company. And building a company isn't the only way to bring your product idea into the world. In fact, it's probably the more difficult route. Licensing your product idea can be a much more lucrative and effective way to go about the process. Licensing involves "renting" your ideas to companies that pay you a royalty on every sale made. Stephen Key, serial inventor and creator of recognizable names such as Laser Tag and Teddy Ruxpin, wrote *One Simple Idea* (McGraw-Hill, 2011) to share the lessons he learned in his 30-plus years of inventing and licensing. Key has successfully separated the creative and (sometimes) lucrative aspects of product creation with the stress and headaches of running a business by finding the point of leverage in the licensing process. His book is the best resource I've found for navigating the process.

Key's formula is straightforward and accessible, but I believe it's getting even easier. With new online platforms and communities, we're entering a golden age for invention. A process so easy, even my mom can do it.

Actually, my mom is no stranger to trying to bring a product idea to life. About five years ago, she, my dad, and a family friend had given the invention process a go. Their idea? A toilet seat with a built in fan. Seriously. I had no idea they were working on it, until I got a phone call from my mom on their way to a meeting with a patent attorney (apparently there was already a working prototype by this point, but I never saw it). Of course, the attorney had gladly taken their deposit to run a preliminary patent search. And guess what? Nothing came up! It wasn't until he ran a more extensive search (and took even more money from my parents) that he realized it was already patented. My folks felt conned.

That's been the process for a hundred years. The road of invention has been rife with hurdles and predatory opportunists ready to take advantage of vulnerable makers.

But now, it's different, as I explained to my mom. To give her an idea of the new maker reality, we decided to walk through the process together using her latest big idea: a steel wool scrubber that was basically a tiny vibrating S.O.S pad. We spent a day making, disassembling, and prototyping her idea, trying to find the shortest route from crazy daydream to "Hey, this might just work." This was a great opportunity for me, too, because I was distilling all the Zero to Maker lessons I had learned into one morning with the least likely maker of them all: my mom.

By the end of the morning, we took apart a Sonicare toothbrush as well as an electric nail polish remover and rebuilt them with new steel wool heads. We tried to clean off dirty pots and pans, then went back to refining our device. By the end of the morning, we had a hacked electric toothbrush that work surprisingly well.

But we didn't stop there. As I explained to my mom, building and creating is only half of the maker process. The other half, sharing, is equally important. So we took a few photos of our creation, wrote a description of what we had done, and put her idea on Quirky (*http://www.quirky.com*), a social product-development platform. On Quirky, anyone can submit an idea, whether an actual prototype, a CAD rendering, or a back-of-the-napkin sketch. After it is submitted, the Quirky community of over 300,000 members begins to vote on and provide feedback to the idea about everything from form to function. After the initial community vetting process, the Quirky team selects a handful of projects every week to continue on to the development stage, bringing in their product and industrial design expertise. The community stays involved as styling and names are chosen. By the end of the development process, the products are offered for sale through Quirky's online store as well as through their retail partnerships with Bed Bath & Beyond, Target, and over a hundred others.

Within hours of posting our project, we had racked up several votes, received comments, and revised our idea. My mom became entranced with the nearly-instant feedback loop.

Our project wasn't selected for future development, but the entire experience was a success. I had given my mom the maker bug. I showed her how easy it was to take an idea and set the wheels of creation in motion. It flipped the switch for her. After giving her a tour of The Mill, a makerspace near her house in Minnesota, she now has enough information to make (almost) any idea a reality: prototype, share, repeat.

Also, it only cost us a morning of experimentation. We didn't waste any money on patent attorneys or wonder what might have been if we didn't pursue the idea. Between the active community and the team of professional designers, the feedback we got from Quirky was the market research we needed to know our project probably wouldn't work.

If you're harboring a big idea or have a quick fix for a problem or nuisance that you encounter every day, the Quirky route might be right for you. Especially if the prospect of starting and running a company seems daunting, a social development approach can really pay off.

Gary Ross, a graphic designer living in Naperville, Illinois, was one such unsuspecting inventor. After becoming frustrated with his wine glasses breaking in the dishwasher, Gary knew there had to be a solution. He submitted an idea to Quirky for "Tether," a simple, flexible plastic rod that acts as a kickstand for wineglasses in the dishwasher. The community agreed and supported Gary's idea, giving feedback on price and usability. Then, Quirky design and production teams jumped into action, redesigning the product for manufacturing and securing retail partnerships. The product launched in December of 2011, and is currently being sold in Target, Bed Bath & Beyond, and on Amazon.com.

Gary was paid nearly $35,000 in royalties over the first year. If he would have tried to go it alone—paying an industrial designer, dealing with the administrative issues with starting a company, creating production tooling—Gary easily could have been in the hole for $35,000 or much more. And, there would be no certainty that his product would be picked up by major retailers.

If you know where to look (and how to share), it's a golden age to be an inventor.

Protecting Your Idea

So far, I've only talked about ways to share and promote your idea. With making and inventing, though, there's another very important side to the equation: how exactly should I protect my idea?

I'm biased. I think the sharing of the idea is the more rewarding (and ultimately productive) route. Actually getting people to like and adopt (or adapt) and use your product is a monumental challenge. I've seen so many makers come into TechShop with "big" ideas that they won't talk about. They make everyone sign a non-disclosure agreement (NDA) before they will show anyone what they're doing. I think this is short-sighted. At the early stages, the odds of someone ripping off your idea are much smaller than the harsh reality that people might not be interested in the product. Keep in mind that it's those early users who give you the feedback that gets you to the next level.

Eric and I (and many other makers) have taken the exact opposite approach, opting to share our designs as widely and freely as possible. And it has made all the difference. The open-source nature of the project is what gave it life. But this isn't always the best option. Some products and ideas are better served going the traditional route of filing a patent, so it's worth knowing about how to approach them.

Because I haven't dealt with the patent process myself, I've enlisted the help of Andrew Rush of IP in Space. He's a highly qualified patent attorney who also knows the reality of maker and DIY projects. On the DIY Space Exploration website (*http://www.diyspaceexploration.com/*), Andrew listed some of the most important things to think about in terms of patents and how they relate to your project. I thought his framing was excellent, even (and especially) if you plan on making your design open source.[1]

---

1. Stephen Murphey has created a great resource for makers with DIY Space Exploration; it's one of my favorite blogs. The entire thing is filled with useful advice, and the patent interview with Andrew Rush (*http://bit.ly/19vnAsw*) is just one example.

Note

This advice and information is for educational purposes only and should not be considered legal advice. You should consult a licensed attorney to discuss your particular needs.

Here is what Andrew has to say:

The open-source hardware community

Many people in the maker and space communities are building really cool stuff with the idea that they'll figure it out, build it, and then release the plans to the general public so that others can duplicate their designs. The important thing to note about open source is that just because you open-source something, it doesn't mean that it is not infringing on a valid patent. If you design and build an infringing device and then release the plans on the Internet, all those people building those devices are also potentially infringing.

This is called "secondarily liable" for patent infringement and you can be liable for inducing someone else to infringe (even if you had no idea it was patented). To put it another way, you might be liable for not only what you do in your garage, but also what others do in theirs.

This is an important fact because lawyers tend to get involved as soon as there is a lot of money on the table. If the open-source hardware movement gets big enough and is built off of patent infringement, it is entirely possible that legal issues will put a serious damper on an otherwise healthy movement.

Ensuring that your design remains open source

Occasionally, a patent will be awarded for an invention by mistake. Although the patent office does its best to ensure this doesn't happen, occasionally it does. Imagine this scenario: person A invents something, discloses it, gets written up in a small unknown journal or thesis and gets filed in a library somewhere. Five years later, person B invents the same thing and files for a patent. After a thorough search the patent is awarded because they couldn't find the small journal or thesis sitting on a dusty shelf somewhere. Person B's patent is technically invalid because the thing was actually invented five years earlier by person A, but the document is still legally binding.

Why is this an issue? Pretend you're person A and you're using the technology that you invented. Person B comes along and threatens to sue you because you're "infringing." Except you invented it. The only way for person A to invalidate it is through the court system—the really expensive court system.

An issued patent has a presumption of validity and it is difficult (and expensive) to overcome that presumption even if you have the proof.

Long story short: if you invent something and want it to remain open source, publish it as publicly as is legally possible so that the patent office cannot possibly miss it. Establish prior art.

Creating patents from patented or open-source hardware

In the United States, there are currently over 8 million issued, and the vast majority of those (95 percent) are not pioneer technologies. A pioneering patent would be the world's first patent on 3D printing or the Wright brothers patent on powered flight. Most patents are just incremental improvements to already existing technologies. They're a slightly more efficient aileron for a plane, or a 3D printer that prints a little more efficiently. Most of them are just tiny improvements.

If there's an open-source rocket engine design, and you come along and invent an improvement for that rocket engine design, you can patent it. This is where licensing becomes important. When you release a hardware design to the public, you can use a license to stipulate how the technology is used and whether incremental improvements can be patented. For example, you can use a license that requires all technology built using your design to be released under the same license. This effectively halts the ability of someone to patent an incremental improvement based off of your design. Ladyada has a great resource on Open-Source Hardware licenses (*http://www.ladyada.net/library/openhardware/license.html*). Also, by publicly disclosing your design, this also keeps anyone else from being able to file a patent on it.

Officially, you need to worry about patents

On the legal front, if there is a valid patent on an invention and you reproduce that invention without permission from the owner, you have committed infringement. In the United States, there are no "fair-use" exceptions (like satire for copyright-protected material) to patent infringement. So technically, from a strictly legal standpoint, any copying or reproduction is considered infringement. This means in theory that the owner could sue and collect "reasonable royalty," which is the penalty for infringement. Damages get worse (three times worse) if you're found to be "willfully infringing" on the patent, i.e., you knew about the patent and deliberately violated it.

The reality of patent infringement

Does this mean that the patent owner would come after you? This is where it gets tricky, even assuming they know you are violating a patent. Because of the high cost associated with patent litigation, the patent owner (often a business) needs to decide if it's worth the cost to take you to court. Defending even the smallest of infringement cases can cost hundreds of thousands of dollars in legal fees. And the bigger the case, the more expensive it gets. Remember the Apple and Samsung battle?

There are provisions in the law to collect legal fees if you willfully infringe, but most hobbyists and small businesses don't have hundreds of thousands of dollars sitting around. So, there's often no point in trying to collect.

However, they have another option, one that is used extensively by the Recording Industry Association of America (RIAA): send letters to alleged infringers requiring them to pay a fee in order to avoid a legal battle. At the moment though, this option is rarely exercised in the patent arena. However, patent trolls have been known to use it with organizations and not individuals. Because again, individuals tend not to have stacks of cash sitting around. So even if you are infringing, from a business perspective there are several reasons why nothing will happen. Consider it a sort of "Russian Roulette" for patents.

When to file a patent

US patent law has what is called a *novelty requirement*. To get a patent, your device has to be new. The word "new" in the patent world means something different from the word "new" everywhere else. Basically it means that if you invent something, you can't get a patent if it's been invented before, and you can't get a patent if you publicly disclose it and then don't file a patent application for more than a year. You have a one-year grace period from your first public disclosure to enforce your patent rights. If you go to a conference and present your rocket engine design but don't file a patent application on it within a year, that disclosure becomes public.

Before March 16, 2013, the United States was a first-to-invent jurisdiction. Here's how it works: inventor A invents a widget and six months later inventor B invents the exact same widget. Inventor A waits 11 months to file the patent, whereas inventor B files the patent immediately after creating it (thereby filing five months before inventor A). In a legal situation, Inventor A would be considered the inventor of the patent. This is an oversimplification of the process but you get the point.

This "first-to-invent" process, while nice in practice, led to some serious legal headaches because it often became a court battle to prove who had the first "a-ha" moment. The rest of the world took a different approach: "first to file." Basically, in the preceding scenario, inventor A would be out of luck because he or she waited too long to file. This simplifies the patent process (and the legal issues) but it also tends to favor well-funded organizations who have the money and the patent attorneys to apply early and often. This could squeeze out the smaller, independent inventor who wants to be absolutely sure this widget is worth patenting before paying the expense of filing a patent.

Patents as a learning tool

Patents can serve as a great learning tool for building hardware. Essentially, it's a blueprint to building a "nonobvious" invention with the inventor receiving legal protection for disclosing the invention. A patent is written from the perspective that a person having ordinary skill in that technical field is able to read it and then build, make, and use that invention without any undue experimentation.

A propulsion engineer should be able to pick up one of Boeing's patents about their propulsion technology, build, and use it without having to invent a bunch of stuff to successfully do so. That's very powerful because patents only last 20 years from the date the application was filed. After 20 years, that technology becomes public domain. Anybody can use it and exploit it however they wish. If you want to build an Apollo F-1 engine, you can do that since all of the patents have expired (assuming you have the money). Although patents are dry and boring, so is every other instruction manual.

Where to look for expired patents

The first place that you can look is Google Patents (*http://www.google.com/patents*). Google periodically combs through the United States Patent and Trademark Office database for issued patents and published applications. They also do the same thing for the European Patent Office.

Finding expired patents is relatively simple because you can constrain the results it will give you based on the issue date or filing date (for example, patents issued 20 years ago or more from today's date). There is one caveat, though, and it's called a Patent Term Extension. If the issuing Patent Office dragged its feet on the application, the Patent Office will determine how much longer it took then was necessary and add that onto the time limit. If it took 6 months longer than was deemed necessary, the patent would expire 20 years and 6 months past the date it was filed.

Prematurely expired patents

When a patent is issued, you have to pay a nominal issue fee, but there are maintenance fees at 3.5 years, 7.5 years, and 11.5 years. The fees start at $500 and go up every year depending on the size of your organization. If you don't pay them, the patent expires and the technology becomes public domain. One of the most famous expired patents in the space industry is the Canfield Joint. It was issued in the 90s and should still be a valid patent, but the patent holder didn't pay the maintenance fee and it became public domain.

US patent office search

The patent office has a system that's called the Public Patent Application Information Retrieval (PAIR) that reveals the history of the patent. It will let you know whether it has expired, if it's in force, if it's an application, if it's still pending, or if it's been abandoned. New patents and published applications are released every Tuesday and Thursday. Unfortunately, Google only checks periodically, so if you want the most up-to-date listing of patents, you have to go directly to *http://uspto.gov*. It's not as intuitive as Google Patents, and it's harder to get PDFs of an interesting patent out of the system, but it is the most up-to-date source.

9

Making More Makers

Throughout my experiences of both the re-skilling classes and the writing, I couldn't help but feel a slight sense of guilt. It all felt a little too obvious. I kept imagining a conversation with my grandfather, and telling him that I—at 27 years old—had come to the realization that I should start "making" things. That I wanted a career and life that was full of meaningful work that I could touch at the end of every day.

He would have been so confused. For him, that would have been a silly, naive statement. Everyone works with their hands; that was the way to make a living. And if it wasn't the direct source of income, it was still a commonplace part of life to build, maintain, and create the tools you needed around the house. It wasn't a "movement" for their generation, it just was just the way things worked.

I don't blame my father or mother, either. They are each makers in their own right. They learned skills and recipes and resilience from their parents, and they always took the time to explain how different things worked. I don't think the generational drop-off in manual literacy happened because of any specific generation or at any exact moment. It was a slow and subtle slide backward. Driven by convenience and a glut of consumable entertainment, the DIY ethos was slowly replaced by DIFM (Do-It-For-Me). It's taken a growing subculture to point out the problems, reinvent the tools to fit within twenty-first century culture, and articulate the value of taking creative control of technology.

But does it really matter? What happens when a child (or an entire generation) grows up without making?

It's tough to know for sure, but some researchers and educators are starting to connect the dots on the importance of childhood making. One of the leading voices in this field of study is AnnMarie Thomas. AnnMarie is an engineering professor at St. Thomas University in Minneapolis. More important, though, AnnMarie is a maker. She grew up playing in her parents' woodshop, taking things apart and trying to make all her own toys. She went on to study engineering at MIT and Caltech and eventually ended up at her current post at St. Thomas. As a professor of mechanical engineering, she began to notice that more and more of her incoming students, even though they were studying engineering, hadn't built anything themselves or even taken anything apart, for that matter. Concerned about this trend, and with young daughters of her own at that time, she began to look further into the issue. She began conducting interviews with makers she knew, like professional engineers and inventors, asking about their experiences in their formative years. She did extensive research into the childhoods of other famous inventors, looking for the early signs of making. Nearly everywhere she looked, she found it. Great innovations and inventions were almost always correlated with a childhood that had access to tools and making experiences.

Her concerns were only exacerbated by studies and statistics she found. She cites a study by the Nuts, Bolts, and Thingamajigs Foundation of the Fabricators and Manufacturers Association, Intl., which found that 72 percent of the American teenagers they polled had never taken an industrial arts or shop class (myself being one of them).

What does that mean? Does it matter?

AnnMarie hypothesizes in a transcript of her 2010 TED talk (*http://bit.ly/ 19GGSP9*):

> *We can think of this as an experiment: What happens to our culture of innovation if we stop introducing kids to the art of making things? We wouldn't expect a musician to be successful if they were only taught theory and then not handed an instrument until college. The same holds true for making. You'd be surprised at how many engineering students colleges see who have never really built anything.*

> *So why are kids making less? Is there not time in the school day for industrial arts class? Are we afraid to let kids build? In a world where some schools have banned recess as being too fraught with peril it's perhaps unsurprising that the concept of kids working with sharp blades and tools could cause concern. But are kids really so incompetent that we must keep them away from real tools?*

One of my favorite school examples comes from the turn of last century when educator John Dewey founded the Chicago Laboratory School, which had a strong emphasis on learning by doing. Children studied the manual arts at every level of their education. Dewey championed the need for children to be allowed to build real things, with real tools. Thus, when the kids decided they wanted to build a playhouse, they got some advice from teachers and did it themselves. A two-story playhouse, custom furniture, complete with the appropriate building permits, designed and built by children under the age of 14. Jump forward 100 years and we have a generation of kids, many of whom may never be taught how to make things with their own hands. I'm not suggesting that we give a two-year-old a chainsaw (mine still has a plastic tool set), but that we acknowledge that, like playing an instrument, making is a skill which takes years to develop and is best started early.

It might be years until we have the data and metrics to validate the importance of making in childhood, but many of us don't think it's worth waiting. If you spend time at any Maker Faire across the country, the most inspiring sight is all the kids lighting up with curiosity and engagement. With our underwater robot exhibit, we couldn't pull the kids off of the machines. With us there to encourage them to take it apart, we find they actually become more interested in the devices and usually ask us the great questions about materials and construction techniques.

It doesn't take a study or report to convince me that's important. AnnMarie feels the same way. That's is why she, along with Dale Dougherty and others, has worked to create the Maker Education Initiative, a diverse contingent of educators who are working to bring making back into schools. They're not just trying to reintroduce the shop classes that have been lost, but they're also introducing many new digital fabrication and maker design skills.

Bringing making back into middle schools and high schools—whether by preserving shop classes or introducing new maker curricula—is only one aspect of preparing the next generation of makers and inventors. There are numerous other ways that making is taking back its rightful place in childhood. From iPad apps that encourage making to family-centric makerspaces, the landscape for kid makers is rapidly improving. As a parent or role model, it's easier than ever to help amplify the maker spirit inside your child (even as you rekindle the spark yourself).

Setting a Good Example

When I asked George Dyson, author and builder of classic Aleutian kayaks called Baidarkas, what drives a person to want to build one of his kayak kits, he had an interesting response. Over the years, he's met dozens of people who've expressed interest in the kayaks, some of whom have extensive kayak experience, and others who had never been on the water. For years, George tried to find the commonalities between those who actually built the Baidarka and those who just talked about it. In the end, he found that it had little to do with experience or education level, but that everyone who actually finished the project all had some sort of role model who had showed them the joy of making.

His insight seemed straightforward and profound. Maybe it was that simple: we just need to set a good example. To dig deeper, I sought out one of the foremost experts in being a good maker role model: Gever Tulley. Gever spent most of his early career working in technology, specifically software development. He knew he was lucky. He had been able to create an interesting life for himself because of his ability to make things. When he thought back on his life and career, he realized that his current situation could be attributed to specific educational opportunities he was afforded, not just formal education. In fact, he thought, it was the informal learning that had been the most instrumental—the freedom and encouragement to break things, to see how things worked, the space to make his own things.

He worried that todays prepackaged and warning-label-filled childhoods were no longer creating those opportunities, so in 2005 he created the Tinkering School, a summer program designed to teach kids how to explore and build their own things. He was straightforward about his approach, promising to put power tools in the hands of capable eight-year-olds. His approach was unorthodox, but it worked. Hundreds of kids have now passed through the program, and they've built everything from rollercoasters (with more than 100 feet of track) to three-story tree houses.

Of course, that all sounds dangerous, and it actually is! But Gever found that danger could be a tool. He theorized that, through controlled and careful experiments with danger, kids could learn the true value of safety as well as develop the creative confidence they needed to succeed on future projects. Danger was so important (and so vilified), he thought, that it was worth writing an entire book about the idea. The product, *50 Dangerous Things (You Should Let Your Children Do)* (NAL Trade, 2011), was a hit, and his TED talk about the topic has been viewed more than 1.8 million times. Quite an unorthodox idea worth spreading.

His success with the Tinkering School eventually grew beyond just a summer camp. Gever went on to start the Brightworks School in San Francisco. Brightworks encompasses similar ideas to the Tinkering School, just spread over an entire academic year. The goal is the same: to inspire kids with what Gever calls "tenacity." He thinks tenacity is the distinguishing characteristic between an idea and reality, the bridge between conception and inception. And that's hard. Gever explained:

> *The default behavior is to stop working when things get hard. But at a young enough age, you can teach them that working hard is fun.*

When I visited Brightworks to talk to Gever, everything about this mentality was extremely visible. The projects glued to the walls and the entire layout of the high-ceilinged space looked like they had been created by a ten-year-old. And they had! The students actually design and build the environment they want to learn in, so it's not surprising that the design looks more like a tree fort than it does classroom. But the most noticeable sight is the energy of all the kids. The entire environment is designed for engagement and experience, and that's exactly what's happening. The first time I went to Brightworks, one of the students came right up to me to show me his dry ice experiment and warned me not to touch it. This was *not* like my school experience.

When I told Gever about George Dyson's theory regarding maker role models, he agreed. That was the same realization that he had reached. Children whose parents gave them tools to build with, or who involved their kids in home repairs, were the same children who had no fear of failure. He said the parents of Brightworks students fell into two camps: makers who understood the value of childhood tinkering, and parents who wanted to reverse the trend of DIFM and saw the same gaping hole in manual literacy that I did. Regardless of which group the parents came from, or how skilled they were as makers themselves, it was the commitment to setting the example for their kids that mattered.

It Takes a Village (Making a Kid-Friendly Makerspace)

Tara Tiger Brown was no stranger to making. As a technologist, Tara was used to making something from nothing. She had started a women-in-tech group in Los Angeles and was regularly meeting with other women and tinkering with different projects. She was no stranger to the hackerspace idea, either. Her husband had started Crash Space in Los Angeles, one of the early pioneering spaces of its kind. She has spent time at hackerspaces around the world learning the ins and outs of what made a successful space and thriving hacker/maker community.

When it comes to making and building a community, she knew what she was doing. But as soon as she became a mother, she realized she needed to start asking different questions. All of the learning and tinkering she was doing at Crash Space (and even at home) wasn't suited for children and, in many cases, wasn't even a safe environment for them. She started to wonder about her son's future. How was he going to learn? How could he have access to the same unstructured tinkering opportunities that had been so valuable for her? What were other parents doing about this?

She wrote a blog post explaining her thoughts on creating a kid-friendly makerspace, a dedicated space equipped with all the trappings of a traditional makerspace—3D printers, laser cutters, etc.—but with all of the safety precautions of a kindergarten classroom. It wasn't an entirely novel idea. There were disparate resources for kid and family-centric making in the LA area. Museums, for example, were putting together kids' days that involved different types of making. There had even been a few events that included kids at Crash Space, but it was always a challenge to find a venue for these family-friendly activities. Tara theorized that if there were a dedicated space, more events would happen.

Not surprisingly, her blog post resonated with a number of other parents (and non-parents, as well). A small group of them began meeting every two weeks to work on planning and organizing. They also continued to host events at other venues, including one with over 100 people turning out for a "learn to code" event and a liquid nitrogen ice cream social.

The meetings and planning continued. Some people stopped coming, some new folks showed up, and Tara and the core team kept the ball rolling. One of the regular meeting attendees was Sharon Ann Lee, who had been working on the redesign and creative placemaking at the LA Mart in Downtown Los Angeles. In particular, she was focused on developing the eleventh floor of the large building into a creative space. There were plans for a media lab, artists' studios, an industrial design lab, and a co-working space. It seemed like a good potential home for the makerspace the group was working on, so they started hosting their events at the LA Mart.

After months of planning, meeting, and organizing, the group decided to take the project to the next level by improving and moving into the space full-time. So they did what any maker project does when it needs to catalyze a community: they put the project on Kickstarter. Over the month-long campaign, they blew past their initial goal of $15,000, raising more than $34,000. It was enough to buy the laser cutters and 3D printers they wanted to equip the space with, in addition to extra funds to hire staff to supervise during open project time on Saturdays and Sundays.

Now the LA Makerspace is up and running, hosting weekly events and engaging kids and families in making projects of all shapes and sizes. There is a kid-friendly makerspace where there previously hadn't been.

For those of you who are interested in creating a similar type of makerspace, here are some tips and suggestions that Tara mentioned or alluded to:

Tap the existing maker community

Odds are, you're not alone in your desire to provide your kids with a broader making experience. A great way to find co-conspirators is to tap your local hackerspaces, libraries, museums, and even schools. Send out an email to a mailing list, or set up an initial meetup. Tapping the local maker community is a great way to gauge interest as well as learn what resources might already be available to you.

Have regular planning meetings and be consistent

Tara told me that the process from initial meeting to actually having the space was four to five months, which is a lot faster than most makerspaces come together (due to lots of planning and logistical work that ends up taking time).

The key for their group was the planning meetings. As people became busy with other projects, or new people wanted to join, the consistency of the meetings was the backbone. It became something the community could rely on. Consistency is key.

Make it kid-friendly, not kid-only

The LA Makerspace team was really conscious of not wanting the space to become just a daycare facility. They went to considerable lengths to ensure that the space was kid-friendly, but not kid-only.

One rule: anyone under 13 needs a parent to accompany them. That rule draws a pretty clear line in terms of what the space will become. In the LA team's case, it's created an environment where families are coming to work on projects together, but also avoids the daycare environment that might discourage a maker without kids who wants to work on a specific project.

Make it a team effort

Tara couldn't do it alone. As was discussed in Chapter 4, a makerspace is only as successful as the community that enlivens it. Aside from the administrative work of organizing, the actual class and content creation becomes a tremendous amount of work. It takes a committed and passionate group to make it all work.

One rule: each of the board members of the LA Makerspace is expected to teach a class or be a mentor during open project time on a regular basis. This sets the tone for the entire community that everyone is responsible for making the space excellent. Not just through care of equipment, but through creation of classes and community engagement.

Digital Natives

Do you remember the first time you saw a toddler playing with an iPhone, iPad, or other tablet device? I'm certain you had the same reaction I did: amazement. Watching how naturally they move around the digital surface always shocks me. It's impossible to see that scene and not think about the future—how these children are growing up with computing devices, not as endlessly improving objects to buy, but as extensions of their own capabilities. They're not going to have to get used to being connected to the Internet all the time: that's going to be second nature.

Of course, it's impossible to know how raising such digital natives will turn out (and it's not hard to see some potential risks). But as far as creating more makers goes, this fact of life can't be ignored. If making more makers is really the goal, it should actually be embraced (especially with this new maker world being so DIT-centric). One team, DIY.org (*http://www.diy.org*), is building an app that blends all of this: creativity, making, and sharing.

It's like the Boy or Girl Scouts for the iPad generation.

There are a suite of skills, such as Animator, Bike Mechanic, Instrument Maker, etc., and each contains a series of challenges that the youngster can attempt. After the challenge is completed, a photo is taken of the project and uploaded to their portfolio to be shared with other kid makers.

DIY.org was founded by friends Isaiah Saxon, Zach Klein, Darren Rabinovitch, and Andrew Sliwinski. Each one of them brought something unique to the table: Isaiah and Darren were experienced animators and designers, Zach had started and run a successful web startup, and Andrew was an all-around maker and software developer. After meeting at a conference, the team realized their overlapping passion for helping encourage the next generation of makers and decided to throw in together.

The diversity of interests and passion of the founding team can be seen throughout the app experience. It's a panoply of creativity. They're not just focused on the maker skills we've talked about in this book, either. A look through the blog turns up projects that range from snowshoes made from orange crates to solar-powered cockroach robots; making potato pancakes to building Rube Goldberg machines. And there's a reason they put "nOOb" next to "beekeeping" and "shoe-making." The goal of DIY is to universally recognize creativity.

As Isaiah Saxon explained it to me:

> There's an age window, between five and eight years old, when you're stepping into your identity. You're establishing your creative confidence... or not. If you develop that confidence, you'll be able to try new things. It inspires a courage to face fear. A fear of not knowing, fear of failure, or a fear of not being good enough. After they get a skills badge, my hope is that we've taken that fear away. There's more leverage the earlier you develop these skills.

Their game-like process of attempting challenges and earning skill patches turns creativity and making into a playful exploration of their surroundings. It's fully engaging. And, of course, because they're encouraging kids to share on the Internet, they've made privacy paramount. Instead of posting as themselves, kids choose one of several animal characters to use.

In addition to providing a playful way for kids to develop their creative confidence, DIY.org is also equipping them with the most important maker skill of all: sharing. As each challenge is completed, the young maker uploads the pictures to their portfolio, which can be viewed by their parents and family members as well as serve as inspiration for other kids. In the same vein that makers everywhere are re-skilling themselves, the young DIYers will be fluent in sharing and creating collaboratively with their peers.

The more I listened to Isaiah, Andrew, and the team describe their motivation for building the app, the more I realized how important their work is. They're helping to give the next generation the creative foundation they're going to need to thrive in this new maker world. Perhaps more important, they're giving parents an easy-to-use resource to help them push their youngsters in a maker direction.

My generation grew up with Saturday morning cartoons. My kids will grow up with Saturday morning making.

CAD for Kids

For me, one of the more challenging aspects of re-skilling has been trying to improve my CAD skills. I've taken numerous classes on Autodesk Inventor as well as played around with all of the free tools like Autodesk 123D or SketchUp. At this point, I can pretty safely create an object that I want to 3D print or CNC, but I'm still nowhere close to being a CAD master. I've become adept at shortcuts like finding and modifying a part from Thingiverse rather than speeding through and creating my own design. I stumble around. Parts get made, but it's always a tenuous process. When I watch Eric breeze through a new creation on Inventor, it's very clear to me that I have a lot to learn and many hours of practice ahead of me, which is fine.

For kids, learning CAD doesn't need to be so complicated. Just like kids who learn languages at a young ages have an easier time, so it goes with CAD.

Chris Anderson shared an anecdote about his son learning CAD on Google Plus:

> Today I mentioned to my 10-year-old that our CNC machine would soon be up and running. He asked what a CNC could do, and I said one example would be to carve a battlefield out of stiff foam for Warhammer figures.
>
> That got his attention ;-) He wanted to know how to tell the CNC what to do. I explained a bit about CAD, and showed him Tinkercad, giving the example of one cube that you could stretch and change.
>
> Then I got busy with something else and left him to figure out Tinkercad himself. I came back an hour later and was amazed to see what he'd designed. A ten-year-old. No training. One hour.
>
> The green stuff we're going to CNC out of a sheet of stiff foam. The rest we'll probably 3D print on the MakerBot. It will take a weekend, but this could be our first 100 percent digital craft project.
>
> This is an example of what I talk about in Makers: manufacturing technologies are getting so easy and cheap (even free) that anyone can use them. Kids today can grow up as fluent in CAD as they are in everything else on computers.

Kids can learn this stuff, probably much easier than many of us would give them credit for. Many of them are already thinking spatially, playing computer games like Minecraft,[1] and intuitively understanding three-dimensional design.

Don't hesitate to get them started with one of the programs listed in Chapter 6. They'll probably figure it out before you can.

---

1. John Robb made an interesting and compelling argument about why Minecraft is laying the foundation for future CAD designers (*http://bit.ly/1ao3Jyv*).

Design Education

In a world in which kids' educational and extracurricular activities are increasingly planned for them (and monitored with precision), proposing unstructured creative time doesn't always go over well. Some school administrators seem hardwired to measure standardized test scores at the expense of long-term learning and inspired curiosity.

Proposing a maker-centric curriculum, even as *part* of the school day, can be administrative dynamite. That's getting easier thanks to the growing influence of Maker Faires, but for Emily Pilloton, one of the pioneers of new design-based curriculum, it was an uphill battle.

To her credit, Emily didn't take on a small challenge. After finishing her education in architecture and product design, Emily and her partner Matthew Miller set out to re-invent vocational training. They created Project H Design and the Studio H curriculum, taking the standard shop class model and incorporating the entire pre-production phase: imagination, design, and planning. In her book, *Tell Them I Built This*, Emily explains why:

> The addition of design gives students ownership and pride in what they will produce, and just as important, asks them to think about why their actions are important. Design makes production personal and meaningful, and develops creative problem-solving and exploratory skills that are applicable in any discipline.

It's one thing to talk or write about re-inventing vocational training, but quite another to actually do it. And to do it, Emily and Matt started in one of the most unlikely places: Bertie County, North Carolina. Originally invited by a visionary superintendent, Chip Zullinger, who had admired their Learning Landscape playgrounds, Emily and Matt travelled across the country full of excitement to put their ideas and experience into action. Just before they arrived, however, Zullinger was let go and the relationship with the school board became strained—Emily and Matt were caught in the political crossfire. Despite administrative headwinds, the pair stuck it out, largely because they believed in their model and the students they met in Bertie County.

They pushed forward with their design curriculum that would, in Emily's words, "build public architecture projects for Bertie County—for, with, and by the hands of its students."

They set out on an unknown course, without any set destination. Instead of telling the students what to create, they invited them to imagine. By flipping the mirror, they encouraged students to be vulnerable and uncomfortable. Emily wanted to show them that you can't just show up with a bag of tools and start building—you need to plan. It introduces a certain rigor to the creative process, and exposes kids to "the non-linear chaos that comes with creativity."

With the Bertie County students, the tangible goal was to create a farmers' market for their town, which they did (and it is beautiful). The underlying goal of Studio H, though, is to create a sense of possibility within the students, a willingness to try and to be okay with failure. By teaching the design process in addition to the tools, they've given the students a real skill for the twenty-first century:

> For our students, the table saw, X-Acto blade, and laser cutter are of equal importance and essentially serve the same function: to cut. We do not teach the tool; we teach the thinking. The tools are a way to achieve an ultimate goal. That goal, too, must always be rooted in citizenship. We must use our tools for the benefit of others. What is their value if not to construct the world we want to live in?

The Makerspace Playbook

Emily's work with Studio H prototyped a new way to think about design and making education. More important, it's had a profound impact on the group of students who have been through her program. The question then becomes, how does that scale? How can Emily's experience translate to your local school district?

Dale Dougherty wondered the same thing. As the publisher of *Make:* and the organizer of Maker Faire, it didn't take long for Dale to recognize the importance of getting the making philosophy back into an education system that so desperately needs reinventing. After watching kids light up with curiosity and confidence at Maker Faire, and having numerous teachers ask how to bring that excitement back into their classrooms, Dale put together a plan.

He created a platform and database of maker educators at Makerspace.org (*http://makerspace.org*) as a resource for teachers who are looking to integrate making into their curricula. The website also serves as a social network and forum for teachers to share their ideas, results, and strategies with one another, creating a positive feedback loop of successful maker education initiatives. They're in the process of creating project guides, modular sections of curriculum that make it even easier for teachers to pull together a syllabus of maker activities that suits their students' needs and interests.

The group created a Makerspace Playbook (the same resource I mentioned in "Creating a Makerspace" on page 78), a step-by-step guide to getting a makerspace up and running. Whether that's in a high school, or just part of a community project, the guide goes through most of the issues that a group would face when trying to organize a space. It walks through everything, from understanding and articulating the value of making in education, to finding a suitable location to stage and design a makerspace, to sample liability waivers. It provides resources and links to projects to fill a syllabus, while allowing for the freedom to adapt and modify the structure to the specific needs of the students or group. It's detailed down to the recommended number of soldering stations but broad enough to be valuable to a teacher without the financial resources to furnish a space.

Most important, though, the Makerspace Playbook is a point of connection. It's an invitation to join a legion of teachers and educators around the country who are embarking on this mission to bring making into the education system. If there's anything I've learned on my journey (and one thing I hope you take away from this book), it's that DIT is the best way to bring a project to fruition.

It's a resource for any educator or parent, not just design or technology teachers. There are ways to incorporate the best of maker education into all sorts of subjects, from science to history or even just a Saturday morning exploration at home. It also doesn't require that you have a background in making or digital fabrication. It's a process you can explore alongside your students, instilling the just-in-time learning methods espoused in this book.

10

Epilogue

No book is easy to write, but this one had its own, unique challenges that made it particularly difficult. Namely, it was hard to document and explain such a dynamic situation. The maker movement is so alive with excitement and opportunity that topics and tools were changing only weeks after I wrote them down (and then again even weeks after I rewrote them). Over the course of the year it took me to write this, 3D printers and CNC machines have continued to drop in price and have increased their performance many fold. The online communities I mentioned had doubled, tripled, or increased their size by an order of magnitude. MakerBot—one of the original open-source hardware companies—was bought for over $400 million dollars. Maker Faire became an even larger and more wondrous spectacle than the original event I sought to describe. And now, no one blinks an eye when a Kickstarter project breaks the $100,000 mark. The path from idea to prototype to product is shortening every day.

The maker movement has grown up, and as more people like you get involved, the tools will become better and more accessible. In another few years, I suspect the maker environment will be an even more exciting (and more difficult) place to define and describe.

For me, none of it seems more incredible than our own underwater robot project. Of course, I'm very biased, but I've had such a thrilling, front-row seat in witnessing a dream come to life. I often reflect back on that initial conversation between Eric and me in the San Francisco hostel, where we conjured up our wildest ideas for what we wished underwater exploration would become. At the time, they were some *really* crazy ideas, but we didn't know enough to know better.

It was exactly that optimistic ignorance that put us in a position to succeed. Falling back on the only resources available to us, we stumbled into the sweet spot of the maker movement—the converging trends that are democratizing creation.

Accessible and Modular Technology

Eric and I didn't have the resources of a research grant or access to a cutting-edge manufacturing facility; we had a TechShop membership. Given our circumstances, we never even entertained the idea of creating a complex design for the shell of our robot, instead focusing on the easiest tools to use that would be "good enough," which in our case was laser-cut acrylic. This did more than just provide us a cheap and easy way to build our structure, it gave us an easy way to *change* our structure. This became incredibly useful after we began getting feedback from others who had built the device.

We also didn't have the technical experience to design and build our own circuit boards, so we based the OpenROV on the popular maker products that everyone else was using, like Arduino. When we first started, there wasn't actually an off-the-shelf product that could process the digital video we were capturing. It wasn't until the original BeagleBone came out that we could finally afford to include a miniature Linux computer inside the robot (in terms of both cost and space inside the robot). As would happen numerous times with OpenROV, many of our problems were solved with technology that didn't exist when we started.[1] Like with the acrylic plastic, the modular design of the OpenROV makes it possible for us to upgrade the hardware whenever a better component comes along. And that's happening with increasing frequency. That flexibility just isn't possible for a design that comes from complex and specialized tooling.

The initial design constraints have become our biggest advantage in a fast-changing, community-oriented manufacturing environment. You have access to the same flexible tools that we did.

The Power of DIT

It went beyond resource constraints, though. There were also large knowledge gaps. Eric and I, like many makers at the beginning of their projects, couldn't quite grasp the entirety of what needed to be done. Instead of hiding behind what we didn't know, we flaunted our ignorance. We sought out other people and ideas that could help us achieve our goals. In doing so, we accidentally stumbled into the magic formula of DIT development and the tremendous power of collaborative creation.

Less than 8,000 days old, the Internet is still in its infancy. The power of being hyper-connected to one another is still brand new. In fact, we only know a little bit

---

[1]. The original BeagleBone, which we used on the first generation OpenROV, retailed for $89. Now, the BeagleBone Black sells for just $45 and is more powerful.

about it. Projects like Wikipedia and Linux have broken down the mental barriers to understanding the benefits of large-scale, collaborative, and distributed initiatives. As ideas become commonplace in the digital universe, the effects are starting to spill off the screen and into the real world. The maker movement is on the front lines of the spillover.

That doesn't mean it's easy. Every day I have to explain to someone how Open-ROV actually works as a business—that giving away our designs actually makes financial sense. Many of us have been conditioned to be distrusting. We're told that good ideas should be kept secret. There are still strong forces in the world that support those traditional means. Nevertheless, the power of radical collaboration is gaining momentum, and it's gathered enough steam that *enough* people get the big picture.

Will it eventually win out over the traditional models? We'll see. Will it fizzle out as just an overly optimistic ideal? I doubt it. Either way, they can't say we weren't having a lot more fun.

Dreaming Big

The subtitle of the book, *Learn (Just Enough) to Make (Just About) Anything*, is bold. It's a tall claim—too tall, I worried.

It started off as a catchy phrase I used in the Kickstarter campaign for the book, and was a unanimously popular vote (among Kickstarter backers) for potential subtitles. At the time, it made me a little uncomfortable. I thought it might give the wrong impression about how much (and what kind of) information the book actually contained. However, as time went on, as I continued to write the book, and continued to follow and document the maker movement, my worry faded away. I came to truly believe in the possibility of making (just about) anything.

The topics and ideas that I never imagined possible—space travel, new species, hovercrafts—began popping up on Kickstarter, and then turning into tangible realities. I saw communities of collaborators create and launch DIY satellites (and groups form with the intention of creating DIY Rockets). I saw a project on Kickstarter raise over $300,000 to create glowing plants. I saw a replica of the *Back to the Future* DeLorean turned into a hovercraft and take a joy ride on the San Francisco Bay.

I saw wild dreams become a reality. Groups of passionate, excited makers are bringing ideas to life I had written off as impossible. I've since learned better.

Whether these groups achieve their goals remains to be seen, but at least they've re-centered the aim and given themselves more hope for a better future.

Maybe we'll have plants that replace light bulbs, or maybe we won't. Maybe we'll all be exploring the Marianas Trench from our laptops, or maybe it takes ten more years. But, we will get somewhere—we will learn something. Either way, we'll certainly have an interesting story to tell, and at the end of the day, I think that's what we're all looking for: a better story to tell. A narrative with more meaning, more excitement, and more agency.

That's the magic of the maker movement. It's an opportunity to take back the story, to redefine our relationship with technology, and to shape the future in which we want to live. It's an open invitation for everyone to participate—to contribute to the world we're all making together.

Index

Symbols

3D printing
    basic explanation of, 110
    choosing a printer, 118
    design libraries for, 115
    hype surrounding, 124
    off-site services, 119
    sense of empowerment from, 111
    technical explanation of, 117
8bitlit, 153

A

accelerator programs, 158
acknowledgments, viii
additive manufacturing, 117
Adobe Illustrator, 119
American Genesis (Hughes), 107
analog tools, 93, 102, 114
Anderson, Chris, 34, 107, 120, 167
Anderson, Tim, 46
Andon, Alex, 55
Antifragile (Taleb), 105
apprenticeships, 99
ArduCopter, 34
Arduino, 22, 44, 59, 129, 194
ArduPlane, 34
Argonaut Jr. submarine, 80
Artisan's Asylum, 76
artist-to-audience ratio, 140
Au, Jesse Harrington, 112
Autodesk, 14
Autodesk 123D Make, 121
Autodesk Inventor, 113

autonomous flying devices, 34

B

Baichtal, John, 74
BeagleBone, 131, 194
Berkeley Tool Lending Library, 81
BioCurious, 24
Bohm, Harry, 58
bolts, 115
book scanner project, 33
boxes, 123
BoxMaker, 123
Brand, Stewart, 88
Branwyn, Gareth, 44, 71
Brightworks School, 183
Brown, Sam, 154
Brown, Tara Tiger, 183
Build Your Own Underwater Robot and
    other Wet Projects (Bohm and Jensen),
    58
Bukiewicz, Joel, 13, 92, 129
Bunny Chair 3D print, 110

C

CAD (computer-aided design)
    design libraries for, 115
    for younger learners, 188
    importance to digital fabrication, 114
    programs available, 114
CAM (computer-aided manufacturing) soft-
    ware, 127
challenges, appropriate response to, 54, 59–
    61, 85

We'd like to hear your suggestions for improving our indexes. Send email to index@oreilly.com.

197

About the Author

 **David Lang** is the cofounder of OpenROV, a DIY community centered around open-source ocean exploration. He is also the writer of the popular Zero to Maker column on Makezine.com (*http://makezine.com/*), which is a public diary of his headfirst dive into the maker world. As a pioneer in the new hardware startup scene, he organized and facilitated the first-ever Maker Startup Weekend, a weekend-long event that used the rapid prototyping tool chain to prove the immense possibility of the next Industrial Revolution. He was named a TED Fellow in 2013. He lives on a small sailboat in the San Francisco Bay.

Colophon

The text font is ScalaPro, designed by Martin Majoor. The heading fonts are Benton Sans and Glypha—the former font was an adaptation by Tobias Frere-Jones from News Goth, which in turn was designed by Morris Fuller Benton, and the latter font was designed by Adrian Frutiger.

The cover photograph is by Gunther Kirsch, and the cover icons were designed by Nate Van Dyke.

CPSIA information can be obtained at www.ICGtesting.com
Printed in the USA
BVOW03s2204131014

370688BV00003B/4/P